The

iPod & iTunes

Handbook

The Complete Guide
to the Portable Multimedia
Revolution

By Contel Bradford

The iPod & iTunes Handbook: The Complete Guide to the Portable Multimedia Revolution

Copyright © 2008 by Atlantic Publishing Group, Inc.
1405 SW 6th Ave. • Ocala, Florida 34471 • 800-814-1132 • 352-622-1875–Fax
Web site: www.atlantic-pub.com • E-mail: sales@atlantic-pub.com
SAN Number: 268-1250

ISBN-13: 978-1-60138-122-4   ISBN-10: 1-60138-122-0

Library of Congress Cataloging-in-Publication Data

Bradford, Contel, 1978-
  The iPod & iTunes handbook : the complete guide to the portable multimedia revolution / by Contel Bradford.
    p. cm.
  Includes bibliographical references and index.
  ISBN-13: 978-1-60138-122-4 (alk. paper)
  ISBN-10: 1-60138-122-0 (alk. paper)
  1. iPod (Digital music player) 2. iTunes. I. Title. II. Title: iPod and iTunes handbook.

  ML74.4.I48B72 2008
  006.5--dc22
                        2008015804

INTERIOR LAYOUT DESIGN: Vickie Taylor • vtaylor@atlantic-pub.com

Printed in the United States

Printed on Recycled Paper

Acknowledgements

First and foremost, I must thank the Lord for keeping me positively thriving in every sense of the word. I have to thank my Sweety for being in my corner for these eight LONG years. It's all gotta be for something — the good and the bad times. I thank my Ma and Dad for molding me into what I am today and never passing judgment for what I have been. One day, we'll all be okay.

I would especially like to thank Yvonne Perry for being such a willing participant in my project — the first to grant me an interview. This section just wouldn't be complete without mentioning John Pospisil, who I actually contacted on accident. Thanks for leading me to Tech.blorge where all the knowledgeable iPod writers hangout.

I'd like to thank my girl, Donna... you melt my heart every time you tell me how much you like my books. A big thank you to Tee Morris. You made for the most intriguing case study of all. I feel like I have interviewed a true celebrity. I wish you continued success.

Thanks for helping out Christy — lending me your expertise and all that. I hope you enjoy the book.

Last but not least, I'd like to thank everyone else who put up with my annoying e-mails and dedicated their time to complete the interviews and help complete this project.

We recently lost our beloved pet "Bear," who was not only our best and dearest friend but also the "Vice President of Sunshine" here at Atlantic Publishing. He did not receive a salary but worked tirelessly 24 hours a day to please his parents. Bear was a rescue dog that turned around and showered myself, my wife Sherri, his grandparents Jean, Bob and Nancy and every person and animal he met (maybe not rabbits) with friendship and love. He made a lot of people smile every day.

We wanted you to know that a portion of the profits of this book will be donated to The Humane Society of the United States. *–Douglas & Sherri Brown*

The human-animal bond is as old as human history. We cherish our animal companions for their unconditional affection and acceptance. We feel a thrill when we glimpse wild creatures in their natural habitat or in our own backyard.

Unfortunately, the human-animal bond has at times been weakened. Humans have exploited some animal species to the point of extinction.

The Humane Society of the United States makes a difference in the lives of animals here at home and worldwide. The HSUS is dedicated to creating a world where our relationship with animals is guided by compassion. We seek a truly humane society in which animals are respected for their intrinsic value, and where the human-animal bond is strong.

Want to help animals? We have plenty of suggestions. Adopt a pet from a local shelter,

join The Humane Society and be a part of our work to help companion animals and wildlife. You will be funding our educational, legislative, investigative and outreach projects in the U.S. and across the globe.

Or perhaps you'd like to make a memorial donation in honor of a pet, friend or relative? You can through our Kindred Spirits program. And if you'd like to contribute in a more structured way, our Planned Giving Office has suggestions about estate planning, annuities, and even gifts of stock that avoid capital gains taxes.

Maybe you have land that you would like to preserve as a lasting habitat for wildlife. Our Wildlife Land Trust can help you. Perhaps the land you want to share is a backyard— that's enough. Our Urban Wildlife Sanctuary Program will show you how to create a habitat for your wild neighbors.

So you see, it's easy to help animals. And The HSUS is here to help.

THE HUMANE SOCIETY OF THE UNITED STATES®

2100 L Street NW • Washington, DC 20037 • 202-452-1100

www.hsus.org

Table of Contents

Chapter 3: Get Ready, Get Set, iTunes..............43

Chapter 4: iTunes Basic Handling:
What You Need to Know 61

Chapter 12: Options Other than iTunes 221

Chapter 13: Tips & Tricks 243

Introduction

The iPod is small enough to fit in your pocket, comfortable enough to take on a routine run, and packs enough power to amplify a party. This tiny device has not only given music a face-lift, it has restructured the multimedia industry. The iPod is commonly used for recreation, but has recently expanded to the business world and beyond with the device's many capable features.

iPods are the innovative creation of Apple, one of the world's leading suppliers in computer technology. The iPod made its debut in October, 2001, and has since outdone all rivals, including manufacturers of MP3 players, in the digital music industry, amassing more than 70 percent of market sales. With more than 100 million units sold, the iPod has become the number one selling digital music player.

iPod models range from the iPod Classic to the iPod Touch. Though small in size, these portable devices have more power and capability than a high-wattage stereo system or expensive DVD player. iPods have the capacity to hold anywhere from 240 to more than 40,000 songs. Although other companies have worked diligently to outdo the iPod, it still towers over the traditional MP3 player, with features such as extensive photo galleries, the latest music, exclusive music videos, and games. Apple also has used iPod technology to create more advanced products, such as the recently released iPhone.

Because the iPod is so tiny and compact, users may assume the device is susceptible to damage. Several tests have been conducted to rate the iPod's durability. Although these challenges have included dropping the device from certain heights and other ridiculous stunts, the iPod continued to function properly. Even after enduring a washer machine cycle (and drying out for a few hours), the iPod boasted its usual power.

The iPod has won many awards, from "Most Innovative Product" to the "Engineering Excellence" honor. It also ranked fourth in Best Computer Product for 2006.

The iPod's easy-to-use interface and solid design has earned it rave reviews by the industry. To piggyback on Apple's success, several other companies in the multimedia device industry are restructuring many of their products to integrate better with the iPod and iTunes.

In a short time, the iPod has transformed the industry, and changed the way we listen to music, rendering previous devices such as the Walkman virtually obsolete. With more advanced iPod models being released at a frequent rate, older versions have become much more affordable. We have witnessed one of the most unique devices ever created — we have witnessed the "Portable Multimedia Revolution."

Of course, your device will come with a set of instructions. You also will have a "Help" feature that is included with the download of your iTunes software. These are worthy tools to get started, but both are capable of taking you only so far in the world of iPod.

In this book, we will detail Apple's remarkable device, from its inception to the current version. We will reveal the momentous milestones of iTunes and the iPod itself, and expose the astronomical figures these two prosperous components are responsible for creating. We will also shed light on the criticism aimed at Apple Inc. — formerly Apple Computer — and the controversy these multimedia products sparked, including the FairPlay issue and invasion of privacy accusations from the public concerning the iTunes Mini Store. Most importantly, we will educate you about two of the most popular inventions in entertainment, explaining in depth why they have been so successful.

After reading this book, you will be equipped with the tips and tricks needed to make you an iPod pro.

Here are just a few of the many topics covered in this book:

- ♫ The origin of the iPod

- ♫ When and why the iPod was created

- ♫ Different iPod models available

- ♫ Common features in most iPods

- ♫ The origin of iTunes and the iTunes Store

- ♫ How to download and purchase content from the iTunes Store

- ♫ How to play and burn music

- ♫ Creating playlists

- ♫ The world of podcasts

♫ How to "sync" from iTunes to your iPod

♫ Available iPod accessories

♫ Rival MP3 media players

iPod History

How It All Began

Apple's revolution of the portable media world began with the introduction of iTunes, which had the power to transfer music from an audio CD to a compressed digital audio format. iTunes acted as a registry with its ability to store musical data. Some of its other features included CD ripping and duplication capability and the luxury to playback online radio shows.

As the world became digital through video cameras and similar devices, advancements in audio did not necessarily match up. Some audio devices were big, awkward, and difficult to manage. These digital media players were not attractive in presentation or performance, yielding numerous consumer complaints. These complaints were the motivation Apple needed to create its own brand that would far exceed the competition.

Development

Tony Fadell, a former executive of Phillips, had goals of creating an MP3 player with much more storage space and capability than the average portable media player. Fadell established his own company to develop the device and pitched it to more prominent, media-based firms. His idea was thought ludicrous by a few and was rejected by Real Networks. However, Fadell's dream became a reality when Apple noticed potential in the idea and decided to fund the project.

Apple developed the first iPod in less than a year. Needing a clever way to introduce the product to the public, Apple recruited Vinnie Chieco, a flourishing freelance copywriter, to aid in the creative department, and Chieco came up with the name "iPod."

The First iPods

Before the iPod sparked any interest, Apple promoted development of its advanced media player, iTunes. A month after the release of iTunes, Apple put the development for iPods in place.

The iPod was officially introduced to the world on Oct. 3, 2001. The first iPod differed from rival digital audio players by using a hard-disk storage drive instead of CD-ROM interaction and flash memory cards. Apple was intent on making a product that would be manageable in size and energy output, yet possess exceptional features. The initial iPod was customized to operate with only Mac computers, using iTunes software as its organizational registry and media player. It was equipped with FireWire connectivity ports and the signature mechanical wheel for navigation. The original unit possessed a storage capacity of five gigabytes — enough disk space to hold up to 1,000 songs.

The first iPod was priced at $399. Even with limited storage space by today's standards, 125,000 iPod units were sold in just over two months following the release.

The original iPod advanced quickly. In March 2002, 10-gigabyte upgrades became available via the Apple Web site offering users an additional 1,000 songs (for a total of 2,000) at the price of $499.

iPod Evolution

The Second Generation

The iPod took a giant leap in the way of compatibility in October of 2002 when the device became available to PC users. While FireWire was still a required method of connectivity, MusicMatch became the management registry and media player software rather than iTunes. These devices were much more durable than previous PC versions because of the trademarked touch-sensitive scroll wheel. Numerous models were released with five, 10, and 20 gigabyte hard drives. The PC models were similar to the first generation in body style, but with much more rounded corners.

By the latter part of 2002, the iPod had grown in popularity and demand. Retailers such as Target, Best Buy, and Dell had picked up on the trend and started selling the device in their stores. The iPod also gained a sense of fame with celebrity endorsements — limited editions were released with engraved signatures of pop queen Madonna, pro skateboard legend Tony Hawk, and music sensation Beck. Another model was released engraved with the logo of the pop band No Doubt.

The Third Generation

The iTunes Music Store was introduced in April of 2003, boosting notoriety and sales of the iPod. Apple mastered e-commerce by opening its advanced online record store. Prices ran 99 cents per song and $9.99 per album.

In June of 2003, Apple expanded on the port for an iPod's dock connector. Cables became available to connect to USB ports, encouraging more PC users to buy iPods. Apple followed this improvement by announcing that it had sold its one millionth iPod.

In September of 2003, Apple introduced the third-generation iPod. This version was much smaller than its predecessor and featured a USB connector port and a touch wheel — a totally non-mechanical feature that made scrolling much easier. These thinner, more stylish iPods boasted larger capacities, storing anywhere from 2,000 to more than 17,000 songs. Third-generation iPods were compatible with Macs and PCs, but the iTunes Music Store was still available only for Mac users. MusicMatch also became obsolete in 2003, making iTunes the sole media player for all iPods.

The Mini Generation

In between the third and fourth generation iPods, the iPod Mini was unveiled in early January of 2004. The Mini was criticized for what many felt was an outlandish price and limited capacity, but Apple eventually increased storage space while retailers dropped the price on the original devices in stock. On the day of the Mini's debut, Apple announced it had reached the sales milestone of two million iPods sold.

In late February of 2004, the iPod Mini began its hostile takeover — iPod lovers crowded retail outlets days in advance. The iPod Mini nearly sold out across the nation in just one day.

The iPod Mini was significant because it aided in the evolution of the device, while keeping the iPod popular at a time when rival flash-drive equipped media players were becoming prominent.

On May 5, 2004, Apple notched its three millionth iPod sale, which signalled the company's supremacy in the market. At the same time, rivals to the iPod began to fade.

The Fourth Generation

Apple made even more advancements with its fourth generation iPod series that debuted in July 2004. The touch wheel was replaced with the click wheel, a concept taken from the iPod Mini. Another quality addition to the fourth generation was extended battery life — these models had enough power to last more than 12 hours. These units were originally priced at $299 with storage space of 20 gigabytes and $399 for 40 gigabytes.

On July 24, 2004, the iPod Mini went global. While the international release was initially pushed back, the impact was tremendous. The iPod was instantly embraced by international fans, selling out in numerous countries.

In August of 2004, Apple declared that iPod total sales had reached 3.7 million. Competition heated up in September and October of 2004. iPods were dominating the multimedia player industry, and their dominance sparked controversy. Microsoft chairman, Bill Gates, openly criticized the popular Apple product. He downplayed the iPod for being a simple device that plagued PC users with viruses. No evidence was ever confirmed that iPod triggered viruses, and Microsoft eventually halted all attacks against Apple and the iPod.

In late October of 2004, the fourth generation received a huge upgrade with the introduction of the iPod photo. Users could now view quality digital images and album cover art. The iPod photo came equipped with a higher number of pixels on an LCD screen, able to display as many as 65,536 bright and brilliant colors. This model supported popular image formats such as JPEG, PNG, and GIF. Perhaps its most amazing feature was the ability to connect to a television to run the slideshow function.

The iPod photo was the most powerful model yet, offering 40 gigabytes of storage space for $499 and an abundant 60 gigabytes for $599.

While nearly identical to the original fourth-generation iPod in appearance, the iPod photo included accessories such as a photo docking station and cable adapters to display digital images on a television set.

The fourth generation also featured more celebrity endorsements such as the U2 special-edition iPod which debuted in October of 2004. This classic model contained engraved signatures from all members of the legendary rock band, plus discounted coupons for U2s music via the iTunes Store.

The Fifth Generation

In October of 2005, not too long after the unveiling of the iPod Nano, Apple introduced the world to its fifth-generation, more commonly known as iPod video. This version included a larger screen and much smaller click wheel. The iPod video model was the first classic style available in black rather than just the traditional white.

Using technology from the previous generation, iPod video also had the power to be connected to a television set or a monitor by use of an adapter. The fifth generation was able to run videos in an MP4 format, with entertainment such as television sitcoms, music videos, podcasts, and movies available for purchase via the iTunes Store. Videos were also able to be downloaded through the Google Video Web site, from which they could be uploaded to the iPod video using iTunes.

Significant upgrades included a screen that was much brighter to enhance video quality, a thorough search feature, gapless playback capability, support for games, and freshly designed earplugs that were more comfortable than the previous earplugs. This update also expanded storage space from 60 to 80 gigabytes, enough to hold up to 20,000 songs.

At this time, iTunes no longer came with the package. Users were now required to download it from the Apple Web site, free of charge.

The Sixth Generation

In the midst of a special iPod event, Apple announced and released its sixth generation of the iPod classic. This model was a bit thinner and much more attractive in appearance — the traditional white color was replaced by a stylish, shiny silver. This iPod was heavily reinforced with a cover plate constructed of anodized aluminum. Battery life was increased drastically, now allowing up to 40 hours for music and 7 hours for video playback. The sixth generation iPod took the portable media experience to another level by offering up to 120 gigabytes of disk space — enough storage to hold up to 40,000 songs.

iPod Numbers

In January of 2004, Hewlett Packard (HP) declared interest in selling its own brand using the iPod name licensed by Apple. Many retailers, such as Wal-Mart, became a part of the merger and began to sell the HP product. Over a year's time, HP iPods made up 5 percent of total iPod sales, but the relationship between HP and Apple came to a crashing halt in July of 2005 due to contractual disputes.

From October of 2004, Apple has ruled the portable multimedia industry, with record sales for a digital music player. In the United States, that translates to 90 percent of market sales for hard-drive media players and just more than 70 percent for media players that use alternative drives.

From January 2005 through January 2006, iPod market shares rocketed from 31 to 65 percent. The number rose to 74 percent later that year.

At the beginning of 2007, Apple reported a company record of $7.1 billion in revenue on a quarterly schedule. These numbers reflect U.S.-based sales, in which iPods accounted for a whopping 48 percent of sales, beating out all other Apple components by far.

Perhaps the most impressive milestone thus far came on April 9, 2007, as Apple announced the sale of its 100 millionth iPod. Later that month, the company reported a second quarterly revenue of $5.2 billion. This was only in the United States, with iPods accounting for 32 percent of those sales. Market analysts believe the popularity of iPods will boost sales of Mac computers and other Apple products.

The iPod continued to make history on September 5, 2007, when Apple hosted "The Beat Goes On" event, announcing that the iPod sales tally had reached 110 million.

In October 2007, Apple reported their quarterly revenue was at $6.2 billion. The iPod made up 26 percent of company sales, Apple desktops made up 19.22 percent, and Apple laptops amassed 31 percent of total sales.

Today, Apple yearly revenue has increased to more than $24 billion in the United States alone and $3.5 billion of this amount is profit. In the most recent fiscal year, Apple profit exceeded $15 billion.

Apple & FairPlay

iTunes and the iPod have had a tremendous impact on the multimedia market. The profit Apple has attained in such a short amount of time is unrivaled, placing the company far ahead in the industry. Apple did not, however, achieve this success without controversy and criticism.

In accordance with copyright laws, Apple was required to implement a system to protect its music and, ultimately, itself. This was done through Fairplay, the company's digital rights management system.

Restrictions were actually in place before FairPlay was introduced. Apple told customers that tracks purchased through the iTunes Store were limited to three computers, with no more than 10 copies of a playlist able to be copied. The system was supposedly tightened with the release of iTunes 4.5 and these restrictions were the result of Apple's revised negotiations with the four prominent record labels.

FairPlay digital rights management system was put in place for protective purposes, but was never 100 percent secure. Since its inception, various applications have been scripted to break through the FairPlay protective wrapper, permitting AAC-formatted files to be used without restriction. An example of this would be a user transforming Apple protected audio files into an unprotected MP3 format. By burning the files to a blank CD, the user could then import the files back into the iTunes software. Although quality declined slightly during the process, the FairPlay wrapper had still been penetrated.

Because a few users violated the terms and conditions, the Apple company came under fire. Most of the finger-pointing came from industry rivals who accused Apple of attempting to monopolize the market. According to them, Apple was intent on locking in current users so they would be exclusive customers of the iPod and the iTunes Store.

No FairPlay — The Attack on DRM

The competition's theory held that the FairPlay combination with iTunes was meant to be rock solid, a factor competing companies attempted to use against Apple. This encryption was exclusive to iTunes, and these music files were not able to be sold on any other Internet music outlet. Other devices that rivaled the iPod could not play these files. That left customers who had purchased content from the iTunes Store no choice but to download their music on the iTunes application and use an iPod to sync it.

On the other hand, the Apple iPod was not originally formatted to play files that were encoded in formats such as Helix, protected by RealNetworks, and WMA, protected by Windows. Unlike its rivals, Apple had a solution to this dilemma — FairPlay. If you had an iPod and purchased music from other online stores, you could play that music with the device by reformatting it to a DRM files.

RealNetworks decided to compete on a similar level in July of 2004. The company introduced its program Harmony, designed to convert files from RealNetworks service into a format compatible with the FairPlay wrapper, which in turn made them iPod compatible. Apple objected, blaming RealNetworks for creating a product capable of hacking into the iPod. Apple retaliated against RealNetworks by releasing a firmware upgrade that made these files incompatible with the iPod.

Controversial times continued for Apple in 2006, this time centering around a French drafting law that focused on protecting content from illegal copying. Clauses of this law would require Apple to provide detailed information on its FairPlay DRM procedures to its competitors. Apple and its supporters strongly protested the law, claiming that it was proposed out of spite for iPod success.

The Breakthrough

Negative opinion persisted regarding Apple's use of DRM. Finally, the company decided to reformat its system. On February 6, 2007, the four major record labels met to discuss new procedures for DRM free music.

At the beginning of April of 2007, Apple and the EMI record label declared that customers of the iTunes Store would be able to download AAC files without the use of FairPlay or any other form of DRM.

iTunes version 7.2 was introduced in May of 2007. This updated edition allowed customers to purchase and download music and other content without DRM attachments — Apple called these new files iTunes Plus. In October, what was once an option became mandatory for those purchasing content labeled iTunes Plus.

Apple received more heat over the addition of its MiniStore feature. This controversial issue will be detailed in Chapter 4.

Case Study: Yvonne M. Perry

Tel: (615) 884-1224

Fax: (309) 402-5236

Web site: http://www.yvonneperry.net

Blog: http://yvonneperry.blogspot.com

I loved to write in my diary from the time I was first able to make complete sentences. I was a senior in high school when my creative writing and poetry teacher, Miss Hallford, praised my writing talent.

It was her encouraging comments that spurred me to continue my writing after graduation. When my children were young, I chronicled their humorous daily events and enjoyed sharing them with my family. When e-mail and the Internet came along, I started writing and sending my short stories to everyone I knew. Many of them encouraged me to compile the stories into a book, which I did. I kept the e-mail format and self-published E-mail Episodes in 2004 — the same year I quit my day job and went full-time with my freelance writing business.

My book *RIGHT TO RECOVER: Winning the Political and Religious Wars over Stem Cell Research in America* was published by Nightengale Press in October 2007. The owner, Valerie Connelly, suggested I enter the book in contests. My book ranked as a finalist in the Current Events: Political/Social category of the national Best Books 2007 Awards by U.S. Book News, an online magazine and review Web site for mainstream and independent publishing houses.

Writers in the Sky Creative Writing Services is a full-service writing and editing company. We work with people all over the globe, offering ghostwriting, editing, and proofreading for any type of article, book, bio, résumé, business document, or marketing text. Our diverse team of six has a wide range of creative abilities including graphic design.

Case Study: Yvonne M. Perry

At first I thought the iPod would be a short-lived gimmick and I did not even raise an eyebrow when it first hit the market. After a year of everyone telling me how wonderful the little device was, I decided I would check into it. Still not convinced it was worth the money, I did not take action. After I started my podcast, I wanted to take my shows with me so I could listen to my interviews in order to critique my interviewing style and improve my technique. My mom, who has always supported my talent and believed in my goals, bought me an iPod for Christmas in 2006. I was tickled pink — the same color as my iPod.

I started my podcast to help people learn more about the book publishing industry. I was answering the same questions for clients again and again, and I thought a podcast would be a great way to disseminate information. A friend of mine introduced me to Audio Acrobat — a great service that allows me to record my interviews through the telephone and access the MP3 files online. The podcast is very popular with writers, authors, publishers, publicists, and readers who want the scoop on a good book. As a compliment and extension to my monthly newsletter, my weekly podcast has brought excellent recognition for my company, my books, and my blog, as well as bringing in clients needing writing and editing.

If something works, why change it? I hope that Apple does not try to fix things that are not broken— unlike Microsoft that changed everything in their 2007 software version. Unnecessary changes like these frustrate users who have to spend hours trying to figure out where a certain feature is located only to learn that it is no longer available in the "improved" version. For example, "auto complete" is a feature I used thousands of times per week in the 2003 version of Word. I no longer have that option in 2007, and now have to type long book titles word for word.

However, I would appreciate having a second button that would skip or advance to the next song without having to go through the menu to find the next selection. This would be very helpful when driving if a song came on that I really did not want to hear at the time or if I tired of the podcast I was listening to, I could just hit the >> next button.

I'm fine with my Nano, but many people love the video capabilities of the newer models. Personally, I would not enjoy watching a video on such a tiny screen. I would love to see iPod compete with Amazon's Kindle device that allows me to store and read thousands of eBooks in a device the same size as a paperback book.

Case Study: Yvonne M. Perry

Yvonne M. Perry — Freelance Writer, Author, and Owner of Writers in the Sky Creative Writing Services

Author of *Right to Recover: Winning the Political and Religious Wars over Stem Cell Research in America,* **www.right2recover.com.**

Platinum Level Expert Author for ezinearticles.com, **http://ezinearticles. com/?expert=Yvonne_Perry.**

Subscribe to Newsletter: **http://www.yvonneperry.net/Writers-in-the-Sky-Newsletter.html.**

Listen to podcast about writing, publishing and book marketing: **http://www. yvonneperry.net/WritersintheSkyPodcast.htm.**

Promote your book or business with our writing packages at: **http://www.yvonneperry. net/Writing_Packages.htm.**

The Basics of Your iPod

Apple has created an amazing device that can do almost anything. Though its design is unique and highly technical, these portable multimedia players are rather easy to work — after you have learned about them.

Since iPod's introduction in 2001, Apple has released a model each year more advanced than the last, testing the limits of technology. Those new to this multimedia craze may view the device like a foreign object, not knowing where to begin. But iPods are generally simple to operate — children as young as eight have learned to master the device. After learning the iPod basics, operating yours will easy.

We will now cover the basics of your iPod from beginning to end. This section will help you become comfortably familiar with the device and help shape you into a master in no time. Topics include:

- ♫   Connecting headphones or ear-buds to your iPod

- ♫   Handling your iPod

♫   Using your click wheel

♫   Navigating iPod display menus and screens

♫   Connecting headphones or ear-buds to your iPod

Your iPod requires an output sound device. You can use the ear-buds included with the package or substitute a compatible brand that will provide equal or better sound quality. Simply insert the small jack of the ear-bud cable into the port for your headphones, which is found on top of the iPod. Your ear-buds should now work properly and produce sound when the iPod is accessed to play audio or video.

Handling Your iPod

When scrolling for music or movies on an iPod, you will mainly use the click wheel as a source of control. To power on the device, simply rotate the click wheel in any direction or press it in any spot. You also can tap the select button on the iPod. The iconic Apple will appear in the center of your iPod screen. The device quickly goes from dark, to dim, to fully powered. When the device is powered on, you will be prompted to select a language preference — this is your first opportunity to work the click wheel. Move your thumb clockwise over the click wheel to navigate your way to the correct language. Once the language has been highlighted, tap the select button. After choosing a default language, you will be taken to the menu screen.

NOTE: It does not require much force to move the click wheel. It may move a bit too freely on the first attempt, but eventually, you will find just the right touch.

Using Your Click Wheel

Now it is time to learn how to make the click wheel move efficiently and follow your lead. When operating an iPod, you are basically going by the choices found on the menu. Simply slide your thumb in a clockwise or counterclockwise direction over the click wheel — this will navigate the menu up or down. As you move through the menu, different items will become highlighted. The trick is learning how to move the click wheel to make it stop where you desire.

After settling on another menu item you wish to explore, simply tap select. This will take you to another menu screen where you can navigate up and down again and choose another option.

In the menu, you will find categories such as Music, Videos, and Albums. You can maneuver the click wheel to access your music. When you have found either a song or album, simply tap the play button in the center of your click wheel to begin the song.

You may wish to backtrack. To do so, tap the menu button at the top of your click wheel to go to the previous screen. Pressing the menu button will move you backward in order of each selected category on the menu.

What makes the click wheel unique is its multifunction ability — it allows you to navigate through the menus and categories of your iPod, and it contains buttons used to access those menus and play songs or videos.

The buttons on your click wheel are easy to find — they are on each 90-degree point around the wheel. All you have to do is lightly press down on the button and the iPod will take action.

Because the click wheel is an apparent circular shape and the buttons are posted on angles, you do not need to be exact when pressing them — tapping down in just the vicinity of the icon will usually prompt the action you are trying to achieve.

When an active song is in process, a "Now Playing" screen will be displayed on the iPod. To regulate the track volume, rotate the click wheel with your thumb — clockwise to increase the volume, counterclockwise to decrease it. Current tracks also can be fast-forwarded or rewound by rotating and pressing the click wheel.

Navigating iPod Display Menus and Screens

With a general idea of how to operate the click wheel and explore the iPod, you are now ready to learn the functions of your menus and screens.

After you turn on your iPod, the screen will display the following menus:

- ♫   Music

- ♫   Photos

- ♫   Extras

- ♫   Settings

- ♫   Shuffle Songs

- ♫   Backlight

When accessing these menus, you will be taken to a screen of submenus. Shuffle Songs and Backlight are menu options that are exceptions; we will cover these areas further on in this section.

As your menu selections guide you to different menus, you should observe an arrow pointed along the right of the iPod screen.

NOTE: The clicking noise you hear indicates the iPod is performing and completing the selected action; if you do not hear this sound, you may not have completely pressed the appropriate button.

Some menus have more submenus than others. When there are more displayable options in the menu, you will find a scrollbar to the right end of your screen. The dark-colored area of the scrollbar indicates how much of the menu has been accessed.

The Music Menu

This is where you begin your jam session. These commands lead to the music menu, which directs you to other related menu options. There are many ways to access music. Here are the options for music on the menu:

Playlists: This takes you to a menu of playlists stored on the iPod. Because you are just starting out, you are likely to have an empty section. We will discuss the creation of a playlist in Chapter 5.

Artists: This command is similar to your playlists. It will take you to another menu in which you will find music categorized by particular artists. Use the click wheel to browse through these artists and find their albums stored on the iPod. From here, you can select either the album or particular song from the artist you would like to hear.

Albums: This category allows you to scroll through your music by albums listed on the iPod.

Songs: By selecting this, you will be directed to another screen with all the audio tracks on your iPod. This menu is listed in alphabetical order.

Podcasts: A podcast is similar to a typical radio broadcasts except that it can be downloaded to your iPod any time you like. Choosing the Podcasts command will take you to a menu that displays all subscriptions you have listed on the iPod.

Genres: This menu is straightforward. By accessing it, you can browse through your music and select by genres such as R&B, hip-hop, and alternative.

Composers: This is where the iPod gets advanced. Songs stored on the device are automatically categorized by the individual who composed the track, rather than the actual performer.

Audiobooks: This menu will display all audiobooks that may be listed on your iPod.

Extras: Accessing this command leads you to another menu featuring extra options on your iPod You can toy with features such as the calendar and clock features.

Settings: This is where you manage preferences for your iPod. Use the click wheel to configure your contrast and color of the screen, adjust settings of the clicker, and so forth.

Shuffle Songs: Different from the other menu items mentioned, the Shuffle Songs command will not take you to a new screen. Instead, this will place your iTunes in shuffle mode, where songs will be played in a random order.

Backlight: This menu gives the option of turning your iPod backlight on or off. Let us go into a bit more details about the functions of the backlight.

The purpose of the backlight is to brighten the screen when you are in areas with limited light — this may be configured to a set time when it will automatically shut itself down. When you wish to use the backlight, slide your thumb over the click wheel and select it. The backlight is now enabled and will instantly brighten your display screen. If a specific time has not been set, then the backlight will typically shut down automatically in a few moments.

Configure Your Backlight

To avoid the hassle of manually powering on the backlight, you may wish to automatically configure it. Here is how:

1.   From your main menu, click "Settings."

2.   From the Settings screen, click "Backlight Timer."

3.   From the "Backlight Timer" screen, choose your timing option: two, five, 10, 15, or 20 seconds.

You may even choose when the backlight comes on. If you want it to power on only when you select from the main menu, you can simply choose "Off" for the timing option.

Your backlight will consume a large amount of energy from the iPod. To optimize battery life of the device, use your backlight only when necessary.

How to Turn Off Your iPod

Powering off your device is as simple as turning it on. Press down on your Play/Pause button and the screen will shut down in a few seconds. To maximize battery life, your iPod will automatically shut itself down if it has not been used in a certain amount of time.

Basic Interface Functions for Most Common Models

Even with all the advancements Apple has made with the iPod, many of its functions have remained routine, making adjusting easier for users who are frequently upgrading. All versions and models come equipped with lengthy white cords attached to headphones. Ear-buds are also included, and the color coincides with the iPod. Newer generations now offer black and white cords and ear-buds. With the exception of the iPod shuffle, all iPods have these five buttons on the front cover:

Menu — This button navigates backward from the Now Playing screen to the Main Menu.

Play/Pause — This button starts or pauses the highlighted song. Pressing play will also turn on the iPod when the lock mechanism is not on.

Previous — This button will skip backward through your tracks.

Next — This button will skip forward through your tracks.

Center — This is the button you will find in the middle of your scroll or click wheel. This selects all items on the menu and enables you to move along freely.

Most iPods have a hold switch at the top on either the left or right side of the device. Sliding the switch to where it shows orange indicates that it is locked — this shuts down the unit and prevents accidental activation, while preserving battery life.

Second generation iPod Minis, fourth and fifth generation iPods, iPod Nanos, and iPod Shuffles have an automatic pausing function when the headphones are detached from the unit.

Difference in iPod Models

While the basis of all iPods is to store and play a large amount of music, each model differs in style, size, and features. Listed below are the iPod models and descriptions that will distinguish them from each another:

iPod Touch — The latest model of iPod allows you to control your music and video experience with a mere wiggle of the finger. A 3.5 diagonal wide screen is a premiere feature of the iPod Touch. This model comes with either eight or 16 gigabytes of storage space. Life of the battery can last up to 22 hours for audio playback and up to five hours for video playback.

The iPod Touch features built-in Wi-Fi, the first of its kind. This model is certainly more advanced than previous versions, with its ability to access the Internet via the Safari browser. The iPod touch also stores your favorite pictures with new and improved editing functions. Cover Flow technology makes for easy navigation while browsing through your music and videos. The iPod Touch also includes an ambient light sensor to adjust the display according to the environment.

iPod mini — Meet the iPod Mini: a smaller, more portable version of the original. This device uses a four gigabyte hard drive to store 1,000 songs with eight hours of battery life. It differs from previous iPods with the inclusion of a Hold switch located on the top left side.

The iPod Mini also spawned a second generation — this model offers disk space of either four or six gigabytes, which stores 1,500 songs. The battery life was extended to give 18 hours of playback.

Delightful color schemes such as silver, pink, blue, or green make the iPod Mini very attractive. Distinguishing the first and second generation of the iPod Mini is as simple as flipping the device over — the hard drive size has been etched by laser at the bottom of the unit.

iPod Nano — Imagine your classic iPod with its comfortable size and great features, and then shrink it down two sizes — you now have the iPod Nano. This version, even smaller than the iPod Mini, is a tiny power packer weighing in at just 7 ounces. The iPod Nano gives two choices — eight gigabytes of storage space for 2,000 songs or four gigabytes for storing 1,000 songs.

Miniature in size yet enormous in capability, the iPod Nano even plays video. Its 2-inch, 320 by 240 resolution screen displays in vibrant color, much brighter than earlier editions. The vivid display screen is great for movies and is terrific for the three built-in video games.

The iPod Nano can be used as your personal organizer, has a calendar feature that allows you to stay up-to-date with your schedule, and has a stopwatch function to keep you steady on your workout regime.

iPod Shuffle — Weighing in at just 1/2 ounce and measuring 2-inches long is the iPod Shuffle. This is the smallest iPod yet. It comes with a built-in clip to fit on your shirt or pants pocket. Its flash drive has one gigabyte of storage space to house 240 of your favorite jams. The iPod Shuffle is skip-resistant, with 12 hours of battery life to keep your music going.

The iPod Shuffle makes managing songs much easier. You can auto-sync your preferred playlists or pick and choose your favorite songs. Play them in your prearranged order or shake things up with the popular shuffle mode. This model allows you to repeat a song in the shuffle, pause, or skip.

The drawbacks of the iPod shuffle include the absence of a display screen and limited space compared with other models.

iPod Video — Often referred to as the fifth-generation iPod, the iPod Video comes fully loaded with power. This model has a hard disk drive with 40 to 80 gigabytes for major storing space. It features a large, wide screen that displays vivid color for video and photos. The iPod Video is available in black or white.

iPod Classic — If 40 gigabytes just will not hold enough songs for you, you have met your match with the iPod Classic. This model carries the most disk space in the iPod array, giving anywhere from 80 to 160 gigabytes — that is up to 40,000 songs. Aside from extensive audio features, the iPod Classic also can store up to 200 hours of your favorite video files for an estimated 100 movies.

Your movies, sitcoms, jamming videos, and pictures will display a real life reflection with the 2.5-inch screen. The easy-to-use click wheel makes navigation a breeze, allowing you to access your music in an instant. Working the iPod classic is simple, allowing you to quickly transfer audio and video from your Mac or PC to the iPod.

The iPod Classic is equipped with a USB port for easy syncing. This model is durable with an enclosure constructed of stainless steel and sturdy, anodized aluminum.

When new iPod models are released, the earlier versions decline drastically in price, making it easy to find a capable device for a reasonable price.

Case Study: Stephanie Graham

Stephanie Graham — Mortgage Trainer

Complete Mortgage Processing

8414 Farm Rd. STE 180-205

Las Vegas, NV 89131

My Podcasts:

http://mortgagetraining.podcastspot.com

www.themortgagetrainingspot.com

My involvement in the mortgage business began in 1991 when the financial institution

Case Study: Stephanie Graham

I worked at decided to save money by cutting its commissioned staff of originators and making this function a responsibility of salaried employees like myself.

I have written seven books for the mortgage industry including *Loan Processor In-A-Box*, *The Contract Processor's Marketing Guide*, and *Slam Dunk: 101 Ways to Win at Loan Origination and Processing*. I also have two books to be released for 2008: *21 Things Every Mortgage Professional Should Know* and *Sharpen Your Credit IQ*.

I became fascinated with the iPod in 2006 after taking Alex Mandossian's Teleseminar Secrets class. Portions of our training class were placed on the iPod for us to listen to. It made me realize that this was not only a great way for me to learn, it was also a great way for me to deliver my own training. Stephen Pierce's Optimization show sold me on podcasts. He has an excellent show with great guest teachers and best of all, it's free! I listen to these shows over and over again.

Podcasting has been effective in my work as a both a marketing and training tool. I marketed a Train-the-Trainer program released in 2007 this way. It is also convenient to offer my student's another means of consuming my training materials.

For me, the best feature of the iPod is the ability to download full feature videos and watch them at my leisure. It has been great for storing videos that have significance in my life both personally and professionally.

I get the most enjoyment out of my iPod when I am traveling or delayed unexpectedly. It allows me to step away from the chaos and confusion of the moment and focus in on what's important to me. I have lots of artists lined up in my iTunes library. I'll list a few by genre:

R&B /Contemporary R&B: Michael Jackson, Eric Benet, Janet Jackson, Alicia Keys, Amel Larrieux, Chaka Khan, Whitney Houston, India Arie

New Age: Jim Brickman, Kitaro, George Winston, Jean Jeanrenaud, Keiko Matsui

Gospel: BeBe and CeCe Winans, Deniece Williams, Yolanda Adams, Cliff Nash

Spiritual/Meditative: James Jacobson, Armand Morin, Abraham-Hicks

Jazz: Michael Ward, James Wall, Kem

Rock: Cindy Lauper, Pat Benatar, Devo, Eurythmics, George Michael

Books & Spoken Word: Dan Kennedy, Roger Hamilton, Earl Nightengale,

Children's: The Cheetah Girls, Disney Hits, Dragon Tales

Easy Listening: Christopher Cross, Celine Dion

Case Study: Stephanie Graham

And the list goes on.

As far as other MP3 players, I have heard of the Zune and Zen but have no personal experience with either. I do not feel that there any true rivals to the iPod at this point.

My favorite model is the video/classic iPod because of its versatility. I use it alone. I use it on a docking station in my meditation room. I attach it to my surround sound system. I play it through my TV. I use it to keep my 6-year-old occupied on long trips. I even have a splitter for the headphones so that whoever I am with can enjoy it, too.

Get Ready, Get Set, iTunes

While the iPod is obviously a phenomenon on its own, the device success would not be complete without the aid of its profitable counterpart, iTunes. What started out as an organizational music database and jukebox quickly has became a multi-billion dollar music store.

The iTunes software allows you to import a large amount of audio files onto your computer. This program is similar to Windows Media Player and MusicMatch. Though it started as an exclusive Apple product, iTunes eventually became available for Windows computers. Its most apparent advantage over Windows Media Player and MusicMatch is integration with the iTunes Store. The ultimate source for music, the iTunes Store also allows you to download videos, movies, TV sitcoms, and audiobooks. This has made the iPod the most powerful portable media player and crowned iTunes the best online digital media store of our time.

Though iTunes integrates beautifully with your iPod, iTunes can be used just as effectively as its own entity. iTunes allows you to listen to the music and watch the movies you download via your computer — all you need is the appropriate drivers and a set of internal or external speakers on your system. Thanks to AirPort Express, Apple's wireless network, iTunes can now be attached to the speakers of your home theater system. This feature allows you to control a home theater system from your computer.

iTunes Milestones Time Line

April 28, 2003 — Apple announced the new iTunes Music Store. It began with an estimated 200,000 songs that could be instantly downloaded, and the buzz was significant, boosting the sales of iPod models while making a reputation of its own. Within the first 18 days of its release, the iTunes Music Store sold an estimated 275,000 songs. This number leaped to a whopping 1 million tracks just five days later.

December 15, 2003 — Apple announced that the iTunes Music Store had sold more than 25 million songs.

April 28, 2004 — The iTunes Music Store continued to make profitable history on its first birthday, eclipsing the 70 million mark for tracks sold. The iTunes Music Store was the only one of its kind with a catalog hosting well over 1 million songs.

June, 2004 — The iTunes Music Store went global. It was now available in the United Kingdom, Germany, and France. The European iTunes Music Store racked up 800,000 song sales within the first week; the United Kingdom alone accounted for more than 50 percent of those sales.

September 1, 2004 — The iTunes Music Store exceeded the 125 million mark for tracks sold.

October 14, 2004 — The iTunes Music Store sales broke the 150 million mark, and nine more countries were included in the iTunes Music Store chain in October: Spain, Portugal, Belgium, Luxembourg, Italy, Austria, the Netherlands, Finland, and Greece.

December 16, 2004 — The iTunes Music Store exceeded the 200 million mark for tracks sold.

March 2, 2005 — The iTunes Music Store exceeded the 300 million mark for tracks sold.

October, 2005 — Apple introduced a new iPod with video playback capability. Videos are sold on the Internet via the iTunes Store, beginning with a collection of 2,000 music videos and TV shows. A few of the shows debuting on the iTunes Music Store were ABC hits "Desperate Housewives" and "LOST." These episodes were able to be purchased and viewed the day after the original broadcast.

September, 2006 — Apple changed the name of the iTunes Music Store to the iTunes Store. Full-length feature films became available. Day-of-release movies sold for $12.99, prior releases sold for $9.99, and users could choose from 75 Disney-owned films from Disney, Touchstone, Pixar, and Miramax.

January 9, 2007 — A catalog of films from Paramount Pictures is stocked into the iTunes Store. The total of films had now exceeded 250 titles, limited to United States customers.

February 12, 2007 — The production company Lionsgate joined the iTunes Store. The total number of films was now at 400.

April 11, 2007 — MGM Studios began offering titles via the iTunes Store, taking the film total to well in excess of 500. MGM Studios offered a variety of classic movies.

August 29, 2007 — A variety of television sitcoms became available in the United Kingdom. Titles included "Desperate Housewives," "LOST," and "Ugly Betty."

Who Can Get iTunes

Apple's mega-music store is available in more than 20 countries, including the United States, the United Kingdom, Canada, Italy, and Japan. The typical method for payment is credit card or iTunes gift card. Users also can purchase from iTunes with PayPal, a free service for sending and receiving payments, along with other similar merchants.

Featured Descriptions

Music — Today, the Apple iTunes Store has more than 6 million songs that can be downloaded for your listening pleasure. These tracks include pertinent information such as the song title, artist who performed the song, what album the song was derived from, and the cover artwork of the album. iTunes 7 now allows users who have not purchased songs directly from the iTunes Store to obtain album artwork. This can be done for free with an account to the iTunes Store. While song lyrics are not included with the tracks, they can be obtained via other Web sites or Dashboard Widgets.

The catalog of the iTunes Store is updated daily with new songs, every Tuesday. The store features a "Single of the Week" that is typically available as a free download for a one week time period.

Video — Initially, the iPod and iTunes thrived off their amazing audio functions. The inevitable capability of video was one the public awaited with great anticipation. The video playback feature was first available with iTunes version 4.8. At the time, the video management area of the software was not necessarily thorough. With limited access to iTunes video content provided by Apple, TV shows and movies were not able to be downloaded by most users outside the United States. Users would have to get their video sources via other Internet sites or from video files on their computer.

Downloading video via iTunes became much more organized with iTunes version 6. This software became available with the launching of the fifth generation (iPod Video). Files were then categorized in sections such Movies, Music Videos, and TV Shows.

Games — Extra games became available with the introduction of iTunes 7. These games are compatible with the iPod Classic and iPod Nano with video playback capability. Games are not able to be played on the iTunes application. Some of the initial games were Cubis, Mini Golf, Pac-Man, Texas Hold 'Em, and Tetris.

Other games recently added include Ms. Pac-Man, iQuiz, and the "LOST" video version. Users are now able to implement their favorite sounds as background music while playing games.

Your Wi-Fi Music Store — This is one of Apple's latest technologies. The iTunes Wi-Fi music store allows you to download music directly via your iPod touch or iPhone. This eliminates the use of a computer and makes accessing your favorite tunes a much quicker, simpler process.

Your Audiobooks — This feature allows you to stay current with your favorite books. The iTunes Store offers more than 20,000 audiobooks that can be downloaded to your computer, then synced to your iPod. This service is provided by **http://www.audible.com**, a Web site that offers more than 40,000 audiobooks. Audiobooks may also be downloaded directly through book Web sites. By using the iPod format, you can transfer an audiobook to your iTunes application and sync it to your iPod. The iTunes Store offers a discount on audiobooks purchased at **http://www.audible.com**.

iTunes Store Subdivisions — The iTunes Store is broken up into subdivisions for easy navigation. Following is a list of these divisions and what you will find in them:

More In Music — In this division, you can search through a number of random stores, such as the iTunes Store or Starbucks Entertainment.

Genres — In this division, you can search through music categories such as R&B, hip-hop, rock, jazz, blues, and reggae.

Pre-orders — In this division, you will find of a list of artists' albums that you can purchase before the release date. This ensures that you will be one of the first to get the latest music.

Celebrity Playlists — The iPod's popularity spans all the way to Hollywood. In this division, you will learn what the stars find hot. Check out playlists created by some of the big celebrities and view descriptions of their favorite songs.

Free Downloads — In this division, iTunes subscribers can download certain songs for free.

Installation

Before you begin your iTunes experience, it is strongly recommended that you check system requirements of your computer for maximum performance. The criteria for software is listed below:

Mac

Operating system X version 10.3.9 or later

Quicktime version 6.5.2

256 MB RAM

Broadband Internet connection (for access to iTunes Store)

Windows

Operating system XP, 2000, or Vista

500-MHz Pentium- class processor or higher

Quicktime 7.1.3 (included with Windows download)

256 MB RAM

Broadband Internet connection (for access to iTunes Store)

Universal Installation

The basic setup of iTunes on a Mac or PC is generally the same, though there are a few issues of compatibility. Let us first detail the basic installation procedure, then move on to slight differences between the two systems.

To install the iTunes application, you need to go to the Apple Web site, **http://www.apple.com**, then select the iTunes tab at the top of the Web page. You then select the latest version of the iTunes application and choose to download it. Before an installation can begin, you must agree to the terms and conditions set in place by Apple.

Just before the installation process, you will be faced with a series of options and selections. You will be prompted with choices to receive frequent notifications from Apple regarding messages on topics such as: "E-mail me new music Tuesday," "Other special iTunes offers," "Keep me up-to-date with Apple news," "Software updates," and "The latest information on Apple products and services." There are boxes that can be checked or unchecked to select or reject these automated notifications.

NOTE: You can omit the process of entering your e-mail address by simply clicking the "Download iTunes" button. You are not actually required to enter this information until creating an account for the iTunes Store.

Users of Internet Explorer will click "Run" to launch the browser installer. In some cases, you may have to run the installer function from your desktop to complete installation. Be sure you have a bit more than 35.6 megabytes of disk space on your computer — this is about how much the iTunes application will claim.

NOTE: Installation of the iTunes program will typically take no longer than five minutes. Users of dial-up will wait longer. This time is usually extended to well over an hour, at times more than two, according to the transfer rate of the Internet provider. Those attempting to access the iTunes Store via a dial-up connection will not have the greatest experience. Though your computer system meets most requirements, a broadband connection is necessary for this reason. Even your faster DSL modem with 512 kbps may take 10 minutes or more to download the iTunes installer to your computer, or it may undergo slow movement in the iTunes Store.

The iTunes installer will begin by prompting you to choose a language. By making a selection, you unpack components of the program, which opens an "installation wizard." After the wizard displays, click "Next" to read the "End-User License Agreement." This agreement is important, as it details the copyright law of the Apple product. It explains how the agreement affects you, the existing penalties for violating it, and other legal matters. If you comply, select that you agree to the terms of the agreement, and then click "Next." You will not be able to proceed without agreeing.

About the Installation of Quicktime

In the next window, you will observe some key information regarding iTunes, such as features and the minimum system requirements you should be working with. From here, click "Next" and instruct iTunes to create a desktop shortcut for you. It will also ask whether you want iTunes to be your default media player and whether Quicktime should be your default video player. If there is another player that has been enabled, remember that your video files will automatically be accessed from there, rather than on iTunes.

NOTE: iTunes and Quicktime may not support all video or audio file types.

If Quicktime is not already in use on your computer, it can automatically be installed with the iTunes software to run video.

From there, you will click "Next" and choose the directory in which you want to install the iTunes application. After choosing a directory, simply click "Next" once, then again in the following screen to initiate the installation process.

When the installation is complete, it will claim an estimated 100 megabytes of disk space on your hard drive — iTunes will claim about 32.9 megabytes while Quicktime will take an added 66.6 megabytes.

Installing iTunes on a Mac

Because they are both Apple inventions, the iTunes software works well with a Mac computer and getting started is easy:

Installation Instructions

Step 1: Power on your Mac computer, access your preferred Web browser, and type the URL for iTunes Store in the browser bar (**http://www.apple.com**). You will then select the "iTunes" tab located at the top of the page.

Step 2: On the home page for the iTunes Store, locate the version for the Mac operating system you are using and click on the "Download" tab.

Step 3: The download should not take long to complete. When it is done, you need to search for the "mpkg" file on your desktop screen. By clicking this icon, you initiate the installation process.

Step 4: Read and follow the prompts and instructions, such as choosing the directory iTunes will be installed on.

Step 5: When the installation is complete, you will be prompted to close all open applications and restart your system.

Step 6: When your system reboots, locate the iTunes icon on your desktop screen. Click the icon to open your iTunes program.

Install iTunes on Windows

Unfortunately, some Windows users run into complications when attempting to install iTunes. Since the introduction of PC iPod versions, compatibility issues have arisen. The most recent problems have come with the Windows Vista operating system, one of Microsoft's latest releases. At the same time, many users also have had successful installations of iTunes via Windows. A few things you need to know before beginning follow.

You will begin by shutting down all program windows. It is important to disable all virus protection and firewall programs on your computer — these applications may recognize iTunes as an error and prevent installation.

Many interruptions with Windows and iTunes installations come with the direct download from the Apple Web site. In this instance, you will need the following steps:

Installation Instructions

Step 1: Download the iTunes setup file from the Apple Web site.

Step 2: Find the icon for the iTunes setup and click it.

Step 3: Click "Next" in the new window to initiate iTunes installation.

Step 4: You will now be prompted with questions for the setup. Make your desired selections and click "Next."

Step 5: The new screen gives these options:

Install Desktop Shortcuts: Allows you to create easily accessible icons for iTunes on your desktop.

Use as audio files default player: Enabling this makes iTunes your primary media player for audio. When you place a CD into your drive, iTunes will be the source to detect the audio and play it back.

Use Quicktime as default player for media files: Enabling this makes Quicktime your primary media player for audio and video.

If you are totally satisfied with your current media player, these options can be omitted. After making your selections click "Next."

Step 6: You will then be prompted to a Choose Destination Location screen. Select a destination, such as your C or D drive — the iTunes installer will typically assume the C drive by default. The installation process begins by clicking "Next."

Step 7: The installation should take a few minutes. When the message displays, click "Done."

iTunes has now been installed on your Windows operating system.

Using iTunes on Your Computer

After installing iTunes software on your Mac or Windows operating system, access it and give it a try. Locate the shortcut icon you created during the setup process and click it — this will display the Apple License Agreement. After reading the terms, agree or decline. By declining, you will not be able to access the iTunes program. If you agree with the terms, you have the choice to save the terms as a document file or print it for your records.

After agreeing to the Apple License Agreement, a registration screen will be displayed. This gives you options such as registering the product online, by mail, or at a later date. This process can be skipped and is not required.

From there, you move on to the iTunes Setup Assistant Screen. This will walk you through the setup to configure your Internet options. After making those selections, you will click "Next" and be prompted with a window that helps locate your music files. Below are the options.

Add MP3 and AAC files — When the music files for your iTunes and iPod are in AAC (advanced audio coding) format, choose this option. This automatically copies those files into the iTunes library. There may be a need to disable this option if you do not want certain MP3 file formats to be automatically added to iTunes.

Add WMA files (Windows users) — Enable this option if you wish to import Windows Media audio files that are unprotected. These will be converted to WMA and ACC file formats to play smoothly on the iTunes programs.

NOTE: iTunes will not convert protected Windows Media Audio files that have been purchased from other online music stores.

The iTunes Setup Assistant will then give you then option of keeping your music folder organized. To organize your music, you must first add tracks to the iTunes music library. This can be done during the setup process or after. If you choose "Yes," iTunes will comb through the files of music on your registry and automatically place them within the library. "Yes" is the default selection which will sort files into according folders when you modify the song title, album name, or artist name. Though the iTunes program is typically supposed to search for files only in the My Music Folder, it will often scan the entire computer, and may also detect audio files that are a part of video and those related to video games as well — this is another reason why you may wish to manually import your music files into the iTunes program.

Changing Your Default Media Player

We briefly touched on some issues regarding the installation and use of Quicktime. If you initially set Quicktime as your default player, that can be changed during the setup of iTunes. Here are some steps to follow:

Step 1: Select Preference from the menu in your iTunes application.

Step 2: From this menu, select Quicktime Preferences.

Step 3: In the Quicktime preferences, find and select File Types.

Step 4: Double click File Types to expand all the options you have. You will see a list of MPEG files (movie picture experts group), for example. Because the Quicktime has been set as the default player, it is automatically handling these files. To manage these files, double click MPEG.

Step 5: You should now see options that read something like this: MPEG System audio and video files. Disable the check box of the files you do not want your default player to access (Quicktime) and the changes should take effect immediately.

If you are still having difficulty playing your MPEG files via your desired default media player, you may need to actually access that application. There are a few steps to help manage this issue.

♪   Locate the "Tools" menu and select it.

♪   From the Tools menu, locate and select "Options."

♪   From the Options menu, select the "File Type" tab.

You should then notice a check box next to your MPEG files. This area may display a gray color, indicating that this media player is designated to handle MPEG file formats by default. Select the check box beside MPEG to enable all these file types to be assigned by default to this media player, then click "OK."

Once these changes have been made, you may encounter a message the next time you access the Quicktime player. This message will read that some of your file types have now been associated with another media program. If you do not wish to be prompted with this message again, click "Do Not Ask To Before This Check Again" and proceed.

iTunes Updates

From here, the iTunes software will automatically scan the Internet for any updates to the program that may be available. This will occur every time you are online with the iTunes program open and is quite similar to the installation process. More often than not, you are prompted to a screen featuring a new license agreement from Apple. Read this document thoroughly before agreeing, as it may be slightly different from the first one you encountered.

Go to iTunes

The last function of the iTunes Store gives you the option to go directly to the iTunes Store by clicking "Yes" or "No." After clicking "Finish," the setup process is complete. Your iTunes screen will open and you can begin to manage your music. Now you have the ability to import your music, keep it organized, and more importantly, sync it to your iPod. You may even access the iTunes Store at your leisure to find all the great content available.

Case Study: James Walker

James Walker

Freelance Writer

Game Section Editor

Tech.Blorge.com

Well, I was first intrigued by video games when I was around eight years old. My uncle Bucky bought me my first console, a Sega Genesis, and NHL '95. I think it was a two-pronged attack to get me to be both a gamer and a hockey fan. Unfortunately, the hockey part only stuck with me until I was about 16 or so.

I grew up a Sega fanboy through and through. Master System, Genesis, Saturn, Dreamcast, or even their handhelds, the Game Gear and Nomad — if it had Sega on it, odds are I owned it. I even owned a Sega CD-X, which I wish I would have kept since they fetch a pretty penny on eBay now.

Now I am a gamer without a favorite console. I am intrigued by the PlayStation 3's potential in the next year or two, and the Xbox 360 has a ton of great games you can get for the console right now. If, of course, your Xbox 360 does not break like mine has on four separate occasions.

Case Study: James Walker

At Tech.blorge.com, **http://Tech.blorge.com**, I am responsible for delivering the latest news from the gaming industry to the people, with a specific focus on news that somehow directly impacts the "console war" between Sony, Nintendo, and Microsoft.

As far as video game related writing is concerned, I have written for Joychix, Xbox 360 Rally and 1PStart. I also have another Web site that I write for that I just started up called 'BingeGamer.net.' Basically, I will be covering news from developers and publishers, as well as news stories I do not otherwise cover at Blorge.

I am somebody who cannnot operate without some kind of music playing. When I am writing, I have Winamp on. When I am in class, until recently, I have my iPod on. Now it is a Zune. I actually got my first iPod on my birthday in 2001. The iPod had come out the day before, on October 23rd, so you could say I was one of the first to actually own the device. I had the 5GB model, and I used it probably every day from the day I got it until it broke. It was not Apple's fault, though — not too much will survive being thrown out of a car window going 70 mph.

As far as iPod games being worth playing, it really depends on the game. Games like Tetris and Pac-Man I have no problem with, but if the iPod starts getting too complicated for its own good it could be bad. You do not want the same problem cell phone games have.

The iPod Touch, however, has a lot of potential. The screen is big enough that you could emulate a D-Pad and Buttons on either end of the screen, and still have a screen size that is roughly the same as the screen on a Game Boy Micro. Then, not only could you have platformers released for the device, but modders at home could emulate classics like Super Mario Bros. or Sonic. Not that you should modify your iPod Touch, of course.

Single best feature? It certainly is not the video playback. Honestly, who has been dying to see their favorite movie in squinty-eye-o-vision?

If I had to pick a single feature that I would consider the "best"... probably the hold button. A lot of the time I have my iPod in my pocket, along with a million other little things, and with other MP3 players that are considered iPod "knock-offs," you can accidentally change songs with a brush of your finger. With the iPod, you just press the hold button and you are set.

When am I able to get the most enjoyment from my iPod? Probably in class. Yeah, I know. I am a real rebel. But in all seriousness, when I am working on my book report or whatnot, I find that listening to music keeps me focused. Since it would be rude

Case Study: James Walker

to blast Motorhead or David Bowie (eclectic mix, wouldn't you say?) through the whole library, having an iPod (or Zune) handy is nice.

The model I would recommend to someone looking to buy their first iPod would probably be the iPod Nano. It has many of the features you can find in the more expensive models, but without the price tag. That way, if you wind up liking it, you can spend the extra to get the bigger and better models.

If I had to pick any site for downloading iPod games, I would say go to **http:// iPodArcade.com**. They have a good assortment of games (mostly text-based) but, as I said earlier, you do not want games for the iPod to get too complex. The real draw is that they are free — all of them. And as we all know, you cannot beat free.

iTunes Basic Handling: What You Need to Know

iTunes is the powerful multimedia software partially responsible for the success of its equally popular counterpart, the iPod. To date, no other music store online has achieved the success of the Apple iTunes Store.

The actual iTunes application comes with all the features you will generally find in a high-quality media player. What sets it apart from most media software is the number of enhanced functions you will find only with iTunes, with access to the iTunes Store being the biggest advantage. Before going further in depth about purchasing and downloading, let us first detail a few common functions of your iTunes software. Below are some of the basics.

Common Media Player Features

Audio playback: This allows you to play all types of audio files. The most common formats you will find are AAC, AIFF, Apple LosseLess, WAV, MP3, and MP4.

Streaming music: Music streaming is the capability to listen to online music in real-time. While the music is not downloaded to your hard drive, it can typically be accessed and streamed at any time.

Graphics: This allows you to view audio-related graphics from your iTunes program. You are also able to customize cover art for any CDs you choose to burn.

MAC USERS NOTE: Your graphic experience can reach the next level by using Fetch Art. This Mac-based application is one of the best programs you will find to creatively customize your jewel cases with exceptional artwork. After installing this program, simply access your iTunes music library, select the song or album you would like artwork for, then click the Scroll icon from your Fetch Art menu — this will match the artwork exactly to your song or album. Keep in mind that this will only function with music from original CDs or music obtained from the iTunes Store.

Internet radio: Here you can customize presets for popular Internet radio shows, such as those found on Sirius.

Audio track info: Your iTunes software will automatically display all available information when you insert a CD into the drive. This includes the artist name, track name, and the name of the album.

Management and organizational tools: iTunes gives you the option to automatically or manually manage the files in your library.

File type conversions: Your iTunes software will convert all file types for easy integration.

Playlist creator: Compose compilations of your favorite songs with this option.

Burn CDS: iTunes also acts as CD duplication software. Burn your favorite CDs or playlists you have created.

Advanced Features

Auto-sync: This function was created for interaction with your iPod. iTunes can be configured to automatically transfer music from your computer to the iPod when connected. Be careful when performing a syncing of your iPod, as it can also erase everything on the device to reflect your current iTunes library. To avoid this, disable the auto-sync function. We will cover all aspects of syncing your iPod later in this chapter.

Smart playlists: This allows you to set parameters on iTunes for automated playlists. These parameters may consist of genre, date of release, or ratings. When importing a new track, iTunes will detect those parameters, match them up, and automatically add them into the appropriate playlist.

Video: Download your favorite movies, music videos, or TV sitcoms to your iTunes application. You can view them from the program or sync them to your iPod.

Movie trailers: iTunes allows you to stream free trailers of upcoming movies. These clips may not be downloaded to your computer, but can be viewed any time from your iTunes program.

Audiobooks: This allows you to listen to audiobooks in audible format via your iTunes program. Audiobooks also can be imported into the iTunes library. This process is similar to importing a CD. Instead of being automatically loaded into your audiobooks file, they will be categorized as audio files. Taking note of this will help you distinguish them from other audio files you wish to add to a playlist.

Combine CD tracks: This feature enables iTunes to act as your personal mixing machine. It allows you to combine songs to eliminate pauses and make your music flow like one continuous track.

Podcast subscriptions: Stay current with your preferred podcasts via iTunes. After subscribing, you will automatically receive the latest podcasts from the selected broadcaster. Keep reading for the extensive section on the world of podcasts.

Edit ID3 tags: An ID3 tag is data attached to tracks that consists of album cover art, artist name, album title, album release date, and genre. Music you buy from the iTunes Store and get from original CDs will typically have this information attached. These tags can be edited and allow you to include your own info from CDs that may have been burned or music collected from other digital media services.

The iTunes Store: This is the biggest perk of your iTunes software. Browse and purchase the latest online music, movies, music videos, games, and podcasts, then download them directly to your iTunes application. This is the largest media store you will ever encounter.

MiniStore recommendation: iTunes detects the music you are playing or the video you are watching and displays related content available for download.

Breaking Down the iTunes MiniStore

Questions were raised and disputes began with one of iTunes latest features, the MiniStore. The MiniStore was created as user interaction, in a sense. The MiniStore is now a part of the edit menu. It displays a miniature listing of songs that you may like relating to your current library. On the other hand, items in your personal playlists are used to make relevant suggestions toward other items — this works for videos and your audiobooks as well. The MiniStore is a way for users to share what they like. It lists top-ranked songs and albums, as well as new releases. It is essentially what the name suggests — a smaller version of the iTunes Store that can be easily accessed from your iTunes program.

Where the Controversy Began

Throughout your use of the iPod and iTunes applications, you will observe many upgrades in both areas. The update of iTunes 6.0.2. was one the multimedia industry will never forget. This edition contained new features that would automatically repair program errors and help maintain stability. This edition also debuted the iTunes MiniStore.

In the beginning, this new feature was embraced as expected — users simply viewed the MiniStore as another way to stay current with music. A problem arose when a few users learned the actual basis of the iTunes MiniStore. To explain it in simpler terms, every time you select a track when your MiniStore is open, the information about that song is transmitted to the iTunes Store to make recommendations to other users.

Rumors evolved quickly, sparking public concern. Some rumors claimed that your music files were sent to the iTunes Store, or even the Apple company, but instead, they were transmitted to affiliate marketing organizations for personal gain. It was even claimed that Apple user IDs were transmitted.

Apple received major criticism regarding this latest update of the iTunes software. Faithful users became weary, feeling as if they should have been thoroughly warned before using the MiniStore. Many people complained this new feature was an invasion of privacy. Some even claimed that

Apple created this new version of iTunes to act as a Spyware program and purposely collect personal information of its users.

Apple insisted that this information was simply used to make relative recommendations. Apple clarified the functions of the iTunes MiniStore with future editions of iTunes. The company also stressed that the MiniStore could easily be enabled or disabled.

Many people questioned the identities of these supposed third-party information collectors; numerous wondered if they truly existed. While Apple has never been proven in violation, some users are still exercising caution when accessing the iTunes MiniStore. You can protect yourself by disabling the MiniStore. No information can be transmitted when this feature has been toggled off.

The iTunes Library

iTunes has built a solid reputation based on a variety of outstanding features. With every edition, the program upgrades offer better quality and more delightful surprises. Though Apple has used the latest technology to create this software, one of iTunes most amazing attributes comes in a very basic form — the library.

Even with all of the program's luxuries and advanced capabilities, it still performs well at what it was intended to be — a digital media player. Even the most mediocre player contains a library and iTunes is very similar. This is the place where all your audio files are stored. When you want to play music from the iTunes application, simply access the desired tracks from the library.

What separates the iTunes library from the competition is capability. While a few media players may stretch as far as video playback, the iTunes library gives you the works. Aside from storing music, it holds your videos, podcasts, music videos, movies, and video games. During your use of the program, you will find yourself going in and out of the library rather frequently.

Protect Your Library

You are liable to become completely attached to this software, leaving your old player behind in a trail of computer dust. Your library will be plentiful with your favorite music and movies, and iTunes will become your livelihood for entertainment — the ultimate media center.

While enjoying the iTunes program, you should remain realistic. Things happen. Incidents out of your control will occur; this includes system crashing. iTunes is one multimedia player that can be called an instant hit. Yet one issue users of this program have faced since day one has been the gamble of losing it all. If your computer happens to fail, there is a great chance that you can lose everything on it, including the iTunes library you worked so diligently to build. Losing all of the music, movies, and podcasts you purchased from the iTunes Store would be a major letdown. Your money and time are valuable.

For the first few years, users had little support on this issue. You had to purchase, download, and compile the iTunes library with whatever content you desired, but play at your own risk. While you could manually back up your library like the typical file, this called for the use of a system utility to do the job. Another option was to copy the iTunes library into the My Music folder. This is not the most trying task, but many of us simply do not think that far ahead.

Things are different now in the times of modern iTunes. The release of iTunes version 7, Apple's newest edition of the mega-multimedia software, comes with its own variation of a reliable backup tool. This function is capable of backing up the contents of your iTunes library to either a CD or DVD, and this includes all of the music and video purchased and downloaded from the iTunes Store, as well as any content you may have added since the previous backup. Whether you have just a few songs in the library or your entire media collection, using the iTunes backup utility should become a habit.

Importing and Playing Music

After learning the basics and setting up, you will want to take your favorite songs and put them on iTunes. There are a few different routes to take, depending on where you want to import music from. Audio files may be imported from either your hard drive or an actual audio CD.

Import from hard drive: Let us start by accessing the File menu at the top of your iTunes screen. Scroll through this category and locate Import; this gives you access to all files on your computer. Your audio files will typically be stored in the My Music folder. Click this to find the music you are searching for, then open it. You can do a simple drag and drop of those files, and instantly import them into your iTunes program.

Import from a CD: This is an easy process in which iTunes does most of the work for you. Simply insert your preferred CD and the program will take a few seconds to detect it. iTunes will then display track information for the CD, such as artist name, name of the song, and name of the album. If the pertinent information is not available, this may display as unavailable. This may happen if the inserted CD is a copy rather than an original disc.

You have the option of importing the entire disc or select songs. For songs you choose not to import, simply disable the check box beside it. From there, find the import tab in the upper right-hand corner of your screen and click it. In a matter of moments, the music from your CD will be imported and will automatically play on iTunes.

You also can control how the iTunes application will react when a CD is placed in the drive. You can command it to instantly import the CD or automatically play the CD. iTunes even allows you to configure information, such as what bit rate and file format the CD will be imported in.

Rating and Organizing

iTunes offers many ways to organize your library. Music can be labeled by name of the artist, name of the song, name of the album, genre, the date it was added, or by the rating system. The rating system is simple. It allows you to give your tracks a rating from one to five stars. This system comes in handy when performing tasks, such as creating a Smart Playlist. To rate songs, select the desired track from your iTunes library and right click it. This will display a small menu window with a variety of choices. Scroll down a bit and find My Rating. Highlighting this will automatically open another window beside it featuring your rating choices: None, One Star, Two Stars, Three Stars, Four Stars, and Five Stars. Select the rating of your choice and it will be applied to the track.

After choosing a rating, you may also organize your music by the rating. Click the Rating tab at the top of your playlist or current lineup and the tracks will automatically be sorted according to the rating you assigned them. This can be done with any content you have stored on iTunes, including music videos and movies.

Accessing the iTunes Store

After importing the music you own, you may wish to explore what else iTunes has in store. You can expand your music library by surfing the iTunes Store. In the iTunes Store, you can browse and buy your favorite songs, entire albums, music videos, full-length feature movies, TV shows, and audiobooks. iTunes also allows you to stream and preview content at any time. You will be able to download more than two million tracks, nearly 10,000 music videos, thousands of movies, the latest TV shows, nearly 20,000 audiobooks, and more than 65,000 free online podcasts.

How to Create your iTunes Store Account

First, type in the URL for iTunes in your Web browser.

Find and click the "Sign In" tab at the top of the screen.

Click "Create New Account" in the screen that displays.

This will prompt you to the iTunes Store Terms of Agreement. Read the text, and if you choose to comply, click "Agree" to proceed.

The next page will consist of open fields to enter details of your iTunes Store account. You will see fields to enter your e-mail account, create a password, and select a security question.

NOTE: The e-mail address you enter also will act as your user name for your iTunes Store Account. Your e-mail and iTunes Store user name must be identical.

Type in your personal password in the first field, then verify by typing it again in the second field. The password fields are case sensitive — typing in a password with any capital letters will display invalid results.

Next, select a security question and answer it in the following box. Choose a response that will be easy to remember. Click the "Continue" tab.

In the following window, you will be prompted to enter your credit card information. From there, click "Done," and your account will be created. You usually will receive a confirmation e-mail detailing your account information. Print this e-mail for your records and store it in a safe place for reference.

Now that you have created an account, return to the Web site for the iTunes Store and click Sign In. Enter the account details, then click Login. You can speed up the sign-in process by asking the Web site to remember your user ID when using the current computer.

Even without an account to the iTunes Store, you can make purchases by redeeming allowances, gift certificates, and iTunes gift cards by accessing the iTunes Gift Options.

Managing Your iTunes Account

The iTunes Store makes managing your account simple and convenient. At any time, you may add or change credit cards, change your password, view the history of all purchases, and set authorization preferences. To access the Account Management page, open your iTunes Store account and click the appropriate icon at the top right of the page.

Buying iTunes Content

Apple has created the most widely used music store on the Internet. With millions of songs available in the extensive catalog, you will be able to purchase almost any song you can think of. These tracks can then be recorded to a CD, synced to your iPod or rival MP3 player, or played from your iTunes application.

The music store also can be accessed directly from the iTunes program by clicking the icon on the left of the screen. This will take you straight to the Apple Web site and display several media items, including featured artists and albums, new releases, iTunes exclusives, and content that has just been added.

On the left of the screen, you will find the primary navigation display for the iTunes Store. This will allow you to browse certain genres, redeem gift cards and certificates, and perform the search function.

On the right of the screen, you will find listings with the most popular and requested artists, songs, and albums. You can scroll through these categories to find items or index through options, Browse and Search.

Browse function: By accessing this, you will see a listing featuring all music listed in the iTunes Store catalog. These listings are sorted by genre and searching through all of them can be rather time consuming.

Search function: This is a similar function to the one you will find in your iTunes program. Instead of searching through your personal music library, it combs through the entire iTunes Store category. This operation is similar to how a Google search works, capable of returning more than 250 relative results.

Once you have found the desired content, it is time to make a purchase. This can be done in one of two ways: using the 1-Click method or the shopping cart.

1-Click is an Amazon-licensed function used to make online purchases quick and easy. This will be your default purchasing option on the iTunes Store. By clicking on the "Buy Now" tab, your song or album is immediately purchased and the download begins.

The shopping cart can be viewed as a more cost-efficient way to buy music from the iTunes Store. This enables you to purchase your desired content in bulk rather than using a per song method. Albums these days normally have more than 10 songs, so this will help you save money in the long run. Songs are typically 99 cents a piece, and albums are anywhere from $9.99 to $12.99.

The shopping cart feature can be enabled from your iTunes program.

♪   Click on "Store."

♪   Click "Buy Using a Shopping Cart."

♪   Click "OK."

By enabling this, the shopping cart will automatically display underneath items in the iTunes Store.

The iTunes Store also allows you to plan for the future. Maybe you are not ready to purchase songs at this time. Save links to songs you are interested in and save them directly to your computer. These items may easily be dragged and dropped from the iTunes Store into the library of your iTunes program. You can then stream the song previews from your personal iTunes library as a reminder. Tracks can even be sampled before a purchase to make sure you get exactly what you want.

Downloading iTunes Content

One of iTunes biggest perks is the instant purchase and download system. The tracks you buy are loaded into a playlist of purchased music and can then be immediately imported to your iPod. The playlist of your purchased music differs a bit from those you manually create using the iTunes software. Tracks are automatically listed and sorted according to the purchase date. These tracks may be added to, sorted again, or completely removed, depending on your preference.

If your download is disrupted or incomplete, there is a remedy. Find the Advanced menu located at the top of your iTunes page and access the Check For Purchased menu tab to continue or resume a failed download.

Syncing Your iPod from iTunes

Now we have come to one of the most important steps. After you have learned the basics of operating the iPod and how to set up iTunes, it is time to sync content from the iTunes software to your iPod.

Most iPods will come with a set of ear-buds and a USB connection cable. These cables have replaced the big cords previously used to connect devices such as printers and scanners. After identifying the cable, you can find the USB port usually in the front or back of your computer. Once you locate the port, plug one end of the cable into it and the other end into the iPod. The wider end of the cable should be connected to the iPod.

Once both ends of the USB cables are connected to the appropriate ports, your iPod should power on and the screen should become bright.

NOTE: Make sure your iPod is fully charged before attempting a sync. Your USB capable also acts as a battery adapter and helps charge your iPod from the computer.

When the iPod powers on, the Setup Assistant Wizard will display on your screen. If the wizard does not display, make sure your USB cable is securely inserted into the computer. If everything is connected properly, the Setup Assistant Wizard will ask if you have read the iPod safety instructions. Select "Yes" to get started right away.

Next, the Setup Assistant Wizard allows you to personalize your device. You can enter a name to give your iPod an identity, for instance, "Contel's iPod." Whenever you are logged in on your computer's user name, this is the name the iPod and iTunes will refer to.

You also can select whether or not you would like the iTunes program to automatically sync your music, videos, photos, and all content in the library every time you connect the iPod — this is actually the default option. This method is much less of a hassle and the route most users will take. Choosing this option will automatically sync any new tracks or playlists you may have created to the iPod.

If your library has exceeded the storage capacity of the iPod, iTunes will display a message and propose a suggested list of content that will fit onto the device.

The ability to configure your syncing options comes into play once playlists have been composed. Instead of adding the entire iTunes library, you may wish to simply select a specific playlist. Below are few steps on how to make this adjustment.

First, connect iTunes to your iPod via the USB cable.

Then, select the name of your iPod and open the menu.

This screen will display a summary of your iPod. This summary consists of lists of all the content capable of being synced to your iPod.

Select "Music Menu" and you will notice all the options you have for syncing your music files. You can now select which playlists will be included in the automatic sync of your iPod.

NOTE: Using this method of syncing will reflect the changes made in your library. When the new sync is performed, it will remove all content from the iPod that was not configured.

Even though you are syncing to the iPod, your iTunes application is the actual headquarters. When you are tired of hearing a certain song or looking to free up storage space, you will need to remove those files from iTunes. When performing your next sync, the removal will take effect.

After finally choosing your method for syncing and understanding the configurations, click Finish and allow iTunes and your iPod to begin the magic. Syncing will take a few minutes. Do not disconnect the USB cable during this process — this will disrupt the sync and can cause considerable damage to your iPod. You will know the process has been completed when the Setup Assistant Wizard displays a message that it is all right to remove the cable from the iPod.

You will then be prompted to register the device. This can be done right away or later, if you choose.

Sync in Manual Mode

Syncing your iPod manually is a great method when you are looking to import specific content. This mode causes your iPod to act as its own library, eliminating the need for interaction with iTunes application. Setting your device to manual mode gives you more creative control over the iPod and limits restrictions you may have with an automatic sync.

From the first time you sync the iPod, it becomes connected with your iTunes library. Because the automatic mode works with a single library, managing the iPod without complication may become somewhat of a challenge. This also can cause difficulty when you want to sync music from another computer. The reason behind this goes back to the iPod's close relationship with iTunes. When attempting to sync content from another computer to your iPod, the device will automatically inform you that it is currently in sync with another computer and library. You will then be prompted to delete the content on your iPod and replace it with contents of the new library.

You may also opt to sync your iPod manually due to the capacity of your iPod. When the device performs an automatic sync, it is importing the actual content of your iTunes library — this includes all of your audio files, all of your movies and music videos, or anything else you may have added. If the disk space on your iPod is limited and you have a huge iTunes library, going with an automatic sync would not be your best bet.

Configuring your iPod to manual mode is not as challenging as it may sound. By following these simple steps, you can avoid the pitfalls of automatic syncing:

1. First, you need to connect your iPod to your computer via the USB port.

2. The iPod icon will then be displayed underneath Devices. Click the icon.

3. This will bring up another menu. Locate the Summary tab and click on it.

4. You will then see the options Manually Manage Music and Manually Manage Music and Video. Select the content you wish to manage.

5. Now all you have to do is drag and drop whatever content you want synced onto the iPod.

Once you set your iPod to manual mode, the changes will remain in effect. Whenever you connect the device to any computer, you will be able to manually sync from the iTunes library. This makes it simple to manage your personal content and access items on other computers.

Because setting your iPod to manual mode enables it to serve as its own library, certain information will not be transmitted back to your computer, such as ratings and play counts. Even though your iPod now resembles the iTunes library, it cannot transfer audio or video back to the computer.

NOTE: Regardless of your iPod syncing configuration, manual or automatic content such as your photos, games, podcasts, and contact information will always be imported automatically.

Sync Other Devices from iTunes

As products of the Apple company, the iPod and the iTunes software were made to work together. Over time, iTunes has become available for other portable music devices as well. These include multimedia players for Nomad, Archos, and a few other notable companies. Some companies are using the aid of portable device plugins to sync iTunes to other brands of media players. For these devices to have a smooth interaction with iTunes, the music synced must actually be purchased from the iTunes Store. Also, there are now a few other applications besides iTunes that will sync to your iPod.

iTunes supports a number of other popular portable music players, with some limitations — most notably, the inability to play music purchased from the iTunes Store. Supported players include a number of Nomad players from Creative Labs, some players from Rio Audio, Archos, and the Nakamichi SoundSpace 2 device. Other manufacturers may also offer integration using a device plugin. A number of third-party programs have been created to help a user of iTunes to synchronize songs with any music player that can be mounted as an external drive.

Though iTunes is the only official method for synchronizing with the iPod, there are a variety of programs available that allow the iPod to sync with other software players.

Case Study: Arnold Zafara

Arnold Zafara

Freelance Blogger/Writer

Tech.Blorge Senior Writer

I have been writing for two years now as a freelance blogger with emphasis on Tech topics since it keeps me informed about the latest trends and updates about technology and gadgets. I have had an interest in technology ever since I got hooked on video gaming and mobile phones.

My responsibilities as a Senior Writer at Tech.Blorge are mainly to report on the latest digital camera releases. Aside from writing for **Tech.Blorge.com**, I also write for Search Engine Journal (SEO, SEM, Search engine stuff), and for **Rotorblog. com** (Web 2.0, Social networks, online communication). Previously, I worked as a contributor at **CallStyle.com** and **Phone-Guide.com** (mobile phones, Voip, digital media players, and wireless tech). I'm also into SEO stuff, with my travel blog at **http://ourparadisephilippines.com** which is part of a SEO Keyword ranking contest.

Blogging had made me tap my writing skills into something financially beneficial.

Blogging entails a lot of patience.

I took up an interest in iPods since the first day it was launched, since the first generation Shuffle was launched. It took me a couple of hours to master its ins and outs.

In my opinion, the best features of an iPod are sound quality and desktop synchronization feature. I have hundreds of songs in iTunes. "My Chemical Romance" gets the most plays on my iPod.

Features I would like to see in future iPod would be a hybrid of the iPod and iPhone.

iTunes: In-Depth Coverage

Everyone recognizes Bill Gates as the billionaire tycoon behind Microsoft Windows and several other PC-based products. Most of us knew Apple manufactured the Macintosh computer, and that was pretty much the basis of it. Well, since the introduction of iTunes, the iPod, and the iTunes Store, Apple has gained much attention in the way of computer technology, and certainly the portable multimedia industry.

Apple's revolution began with the invention of the iTunes software. What began as a slightly advanced version of the typical multimedia player has quickly evolved into the world's most popular and profitable online music and video store.

In the previous chapters, we learned a bit of history regarding iTunes inception. We learned all the great features of the iTunes software and how those features can be enhanced by the iTunes Store.

After installing the iTunes software, you have learned how to successfully sync your iPod from iTunes and are anxious to build your own content.

It is time for you to master your iTunes program with in-depth training knowledge.

Playing Your Music

This is the place to start with your iTunes application. While the program consists of many features and functions, you are probably looking to play music. There are two basic ways of playing your music in iTunes: from the library or from music stored on a CD.

Playing Music from the iTunes Library

To fully test the capability of the iTunes multimedia software, you must first have music in the library. Unlike several other players, this program will not come with default or sample tracks. Until you import your own tracks, the iTunes library will have a clean slate. We have covered the topics of importing music from your computer hard drive and an actual audio CD, so let us assume you already have a few tracks stored in the library. The simple steps of accessing and playing your music follow.

Open the iTunes application from your desktop or start menu. At the top left, you will notice the Library Tab. Under that, you will see the five categories that make up the content of your iTunes library: Music, Movies, TV Show, Radio, and Podcasts. Click on the Music category.

The full index of your iTunes music library will display in the center of this screen. This will list all of your music stored on the program, typically in alphabetical order. Even if your content is listed as full albums, each individual song will be displayed. Highlight the song you would like to hear, then double click on your mouse. This will instantly begin the track.

There is another method of playing your music from the iTunes library. Just above the "Library" tab at the top left of the page, you will observe a mini media player. This will consist of a rewind and forward button with a play button positioned in between them. Highlight the track you would like to hear and click the play button. Your track will begin instantly.

Also note that the iTunes application will play the tracks of your library in succession. This means you can go from hearing one artist's CD directly to a different one.

Playing Music from an Audio CD

Another way to get audio playback from your iTunes software is by using a real CD. Perhaps your library has reached its capacity of songs, or maybe you would just like to sample a CD before deciding to add it to the library.

Playing from an audio CD is similar to the way in which you import music. All you have to do is insert your disc and allow your iTunes software to do the work. In a matter of seconds, the program should detect the CD you inserted. Track information such as artist name and album may be displayed on the screen, depending on where the CD came from. If the content is not recognized, iTunes will display a message stating so. Your CD will show the proper amount of songs but will be labeled as unknown in the fields for Artist, Track, and Album. The program will ask whether or not you would still like to import the CD. Because we are just giving the music a try, you will select "No" or "Cancel." The track list for the CD will remain on the screen. Highlight and click the track and iTunes will start the song for you. Selecting and clicking the first track will automatically run the entire CD.

Creating a Playlist

Very quickly, you will become quite familiar with your iTunes application. Importing CDs and playing them is the most basic task. Before long, you can have your own mini record shop stored in the library.

After a while, you may experience the need for change. If you are tired of hearing your songs in the same order, it is time to learn the next step — creating a playlist.

Playlists are a great feature of the iTunes software. Playlists allow you to take all your favorite tracks and combine them into the ultimate music compilation, and these playlists have very limited restrictions. The typical music CD contains anywhere from 80 to 120 minutes of audio. With a playlist, you can listen for as long as you want.

To get started with a playlist of your own, search for the "+" button at the lower left side of your iTunes display. Double click the button and a box for "A New Playlist" will branch from that menu — this default text will be highlighted in blue. From here, delete that text and enter a clever name for the momentous musical compilation you plan to create: for example, Jewel Jamz.

NOTE: You also can create a new playlist by accessing the file menu found at the top of your iTunes screen. This should be the first option within the menu. By clicking this, you will be prompted to enter the name of your playlist.

Once you have the given the playlist a name, it will instantly appear on the left-hand of the iTunes screen under the "Playlists" text. Now it is time to fill your playlist with songs.

At the top of the same column, you will notice a "Music" tab just underneath the "Library" text. Clicking on "Music" will display a full-length scroll of your iTunes library. As if you are about to play a song, highlight the track you would like added to your playlist. After right clicking it, another menu will be displayed with options for the selected track. The items you will find are:

♫ Get menu

♫ My rating

♫ Show in Internet Explorer

♫ Reset play count

♫ Reset skip count

♫ Convert selection to AAC

♫ Play next in party shuffle

♫ Add to party shuffle

♫ Uncheck selection

♫ Add to playlist

♫ Show in playlist

♫ Copy

♫ Delete

♫ Assort field

When stopping your cursor on "Add playlist," a window will automatically show on the right side of the menu. This dialog box will show an index of all the playlists stored in your iTunes library. Because we have made only one playlist, this is the one to choose. The song is then instantly added to the playlist. Repeat this process with each song you wish to add — this can be done at any time. Your playlists can always be built upon.

Smart Playlists

You have now compiled enough music to play for days. The playlists feature is actually basic in most digital media players, but what makes iTunes different is the ability to customize your playlists. The program also enables you to expand on this with the introduction of smart playlists.

So What Exactly Is a Smart Playlist?

A smart playlist is simply iTunes' method of helping you create different types of music compilations. The program makes suggestions and helps you create playlists by genre or trend. If you wanted to create a playlist strictly from a list of your favorite love songs, iTunes will automatically create a playlist for you that is based on your preference. To see how this is done, follow these steps:

Locate the file menu at the top left corner of your iTunes screen.

When the index of that menu displays, scroll down and select "New Smart Playlist."

This will bring up the "New Smart Playlist" dialog box. You will then be prompted to "Match the following rule." This is how iTunes helps you select the exact criteria you want the playlist to be composed of. The first box will be based on categories such as artist, album, compilation, composer, and date added. iTunes even allows you create smart playlists based on what type of format your file is.

The next available section allows you to instruct the iTunes application of what type of tracks to look for. You have the choices of "Contain," "Does not contain," "Is," "Is not," "Starts with," and "Ends with."

The last field of that row is where you type in the criteria you want the smartlist to be based on — R&B, Beyonce, or WMA, for example.

The smart playlist function also allows you to create by the number of tracks or the duration, among other categories. With the iTunes software, you even can enable the "Live update" function to ensure that your smart playlist remains current.

The iMix

Part of what makes iTunes stand high above the competition is the method of advanced user interaction iTunes employs. Sharing your playlist with others is made possible by the iMix feature.

The iMix was introduced with the iTunes version 4.5 update. This allows you to publish your own playlists on the iTunes Store at no cost. These mixes are limited to 100 tracks and the content must be available on the iTunes Store. Once uploaded, iMixes become public and are able to be searched and accessed by any user with an account to the iTunes Store. Your mix will remain on the store up to a year from the original publishing date.

The iMix In-Depth

The iMix works by the matching of audio tracks on the iTunes Store — if the song was purchased from the store, it should match up fine. Tracks can also be matched through the iTunes search and filtering system. In this instance, the match relies on similarities of the tracks in your library and those found in the iTunes Store.

Sometimes tracks may not work because a match is not possible. This typically happens when file formats are not relative or the song has come from a source outside of the iTunes Store. The best way to make sure your tracks will match is to preview them in the iTunes Store.

Rip Your CDs with iTunes

The practice of "ripping" came along with the CD burning era. This is actually the method in which a CD is recorded onto your computer. Ripping is when songs are pulled or "ripped" from a CD, then stored onto a computer. When iTunes imports the data from your CD, it is essentially performing the ripping process. Below are a few steps that will show you how to rip your CDs directly to the highly compatible MP3 format.

In the top left corner of your iTunes screen, locate and click the "Edit" tab.

Click on "Preference" from the edit menu.

Scroll down from here and click the "Advanced" tab.

In the "Advanced" menu, click the "Importing" tab.

Click "Import using," then select the "MP3 format encoder."

Click "OK" to complete the process.

This is a neat trick that enables every CD ripped from iTunes to transfer in MP3 format. To change this configuration, follow the steps above and make the needed adjustments in the "Import using" area.

Burn Your CDs and Playlists

While learning the basics and in-depth procedures of iTunes, you will likely fill your library with all of your best music. Another of the many great features of iTunes is CD burning capability. You can burn all the classic tracks from your playlists or the full versions of your favorite CDs. While the functions for burning are basically the same, there are a few different ways to go about it.

Simple Method

Select the playlist you wish to burn under the "Playlist" text in the right-hand corner of your screen.

Open the file menu at the top left corner of your iTunes screen.

Scroll down the menu, find and select "Burn playlist to disc." When this is done, the iTunes program will automatically open your disc drive. At the top of the program, in the center of the screen, you will notice an open, light-colored field. The text, "Insert a blank disc," will blink repeatedly across the screen.

Next, insert a blank disc and close the drive. Be sure to allow the drive a few seconds to detect the writable CD. Attempting to start a burn before the CD has been detected usually results in the disc being ejected. This also will give you an error message saying that the disc was not blank. A good way to make sure the CD has been detected is to simply observe the behavior of your disc drive. You will notice a green or red light blinking that indicates that the software is the process of detecting the disc. Once the CD has been recognized, simply click the "Burn" icon located at the bottom right corner of the iTunes screen.

At the top of the screen, you will observe the status of the current burning process. The iTunes software will indicate what song is burning, at what speed, and the estimated time of completion.

There also is a much simpler method of burning your playlist to an audio CD. By accessing your playlist, all of its contents will be displayed on the iTunes screen. Simply locate the "Burn" icon.

Advanced Method iTunes CD Burning

iTunes is very advanced software, yet very easy to use with a bit of practice. The program gives many ways to perform similar functions. Burning your playlists and audio CDs can be done in a few different ways. The steps below show a few different ways to customize the CD burning process:

Select the "Edit" tab in the top left corner of your iTunes screen.

From the edit menu click the "Preferences" tab.

In this dialog box, locate and click the "Advanced" tab.

Now check the box, "Burning."

You will notice there are a few extra options for burning just beneath it: Audio CD, MP3 CD, and Data CD or DVD. Read below to learn the functions of these burning methods.

Audio CD: Relates to the typical audio CD you will attempt to burn. The iTunes program uses technology that makes your recorded CD compatible with almost any playback device. This is a standard method, so it is formatted to meet the capacity of blank media typically no larger than 702 megabytes.

MP3 CD: Here is one burning option you should consider. This allows iTunes to duplicate the music files in MP3 format rather than audio tracks. By doing so, the files are compressed and able to fit on MP3-compatible blank media. This makes your finished product accessible with other popular MP3 devices.

Once you have chosen a method to burn, it is time to view the results. After inserting a disc, simply locate the "Burn" button in the lower right hand corner and click it.

NOTE: The burning process will generally take a few minutes to complete. This depends on the actual duration and file format of the song. Also, although you have the luxury of burning your custom playlists to a CD, there is a limitation. This is based on the capacity of your blank CD media. To simplify things, you may wish to divide your ultimate compilation into a few that will meet the amount of disk space the CD is working with.

Learn the iTunes Shuffle

The iTunes software gives you several great ways to enjoy your music; consider it your personal jukebox in digital form. The "Shuffle" or "Party Shuffle" function allows you to mix things up and keep the party going. Set up the music in any order you like or add the element of surprise by customizing your jams to a random setting. While it is essentially in the form of a playlist, shuffles differ in the way they operate. Let us now go into the basics of how this amazing feature works.

Your Party Shuffle feature on iTunes is a more dynamic version of a playlist. The concept was actually spawned from the iPod Shuffle model, with which it shares many similarities. The exception is that you have your computer screen to look at. The Party Shuffle mode will shake up the songs in your iTunes library. If you do not like the song that is playing, simply click "Next" on the mini media player, located at the top of your screen, to advance the track. If you choose not to hear a song selected in the playlist, simply check the box beside it and that removes the track from your shuffle. iTunes will automatically present you with other songs to include in the mix.

Here is how to set up the Shuffle feature on iTunes:

On the left side of the screen, you will notice the "Shuffle" or "Party Shuffle" tab. Click it.

The Party Shuffle will open up in its own window. In this menu, you will find random tracks displayed on the screen. Toward the bottom of the menu, you will notice the configuration functions, Source and Display. These let you adjust the source from which the content will be pulled and how the play order will be displayed. The source you select will be from playlists you may have created or individual tracks in the iTunes library. Choose your source to proceed.

From here, the iTunes software automatically re-sorts a new lineup from the selected source. If you are not thrilled with the shuffle, iTunes has the power to change it. By clicking the Refresh button at the lower right of the screen, your shuffle is mixed up again and will return new results with a different lineup of songs.

Adjustments to your Party Shuffle can be made even when the mix is in play. In your Party Shuffle window, you will notice the text plays higher-rated songs more often. You can enable this by clicking the check box. For this to take effect, you will usually have to wait for the current shuffle to run or try your luck by clicking the Refresh button.

Similar to a playlist, your Party Shuffle can be built upon and made even better. Simply highlight a track that would make a perfect fit, right click it, then scroll down to Add to Party Shuffle. This places the desired song at the rear of the current Party Shuffle and includes it in the mix. You also can add tracks manually by dragging and dropping them into your Shuffle.

Your job as a DJ has never been easier with use of the iTunes Party Shuffle. Use your finely tuned skills to fire up a social gathering or mix things up at home. This is one feature you are sure to use over and over.

Using the Radio Feature in iTunes

For those of you addicted to sports and talk shows, iTunes has you covered. The radio feature gives you access to some of your favorite radio stations using Streaming Technology.

With iTunes radio, you have unlimited access to hundreds of station spanning the globe. You can choose from all of the most popular categories: R&B, hip-hop, rock, reggae, and more.

Playing with Your iTunes Radio

Learning to work the iTunes radio feature is simple. To find what stations are available, click the "Radio" tab on the left side of your iTunes display screen. This will unveil an index of categories. Click on your desired selection to explore the content inside the category. All radio stations will be represented by radio icons located to the left side of the text. The icon is easily identified, as it sort of resembles a tree at first glance. All available radio stations will have a name, short description of the genre, and information regarding bit rate of the streaming process. For users with high-speed Internet, there is no need to worry. Your music will stream smoothly with no pauses, skips, or interruptions. There also is hope for those with dial-up modems. Because the iTunes radio function requires more bandwidth, you should select a stream with a low bit rate to achieve maximum performance.

Radio stations are frequently updated by the iTunes staff. To ensure you have the most recent list of shows, you will need to refresh the selections. Do this by highlighting the "Radio" text and clicking "Refresh," located in the top right corner of the iTunes display.

Browsing through iTunes' extensive list of radio stations can prove to be time consuming.

Similar to other content of the application, playlists of your favorite radio shows can be organized in a playlist. You initiate this process just as if you are making a playlist of your music. After naming your creation, drag and drop the selections into the radio playlist, and you are done.

Adding Unlisted Radio Channels to iTunes

Because the variety of radio stations on iTunes is not as vast as the music collection, there is a possibility of your preferred station not being listed. If this is the case, then it is time to get advanced.

To enter your own channels into the iTunes application, you need to first acquire the URL for that stream, which will not always reflect the URL from the actual Web site. The best way to identify the correct URL for a radio station is through your iTunes program.

After finding the proper URL for the radio stream, you will need to locate the "Advanced" tab at the top of your iTunes screen, click it, and select "Open Stream" from that menu. The process will initiate and in a matter of moments the radio station should begin to play. The radio station will now be visible in your iTunes library. While iTunes will automatically implement the channel into the library, you may add this to your radio playlist for much easier access.

If you cannot find the URL for the stream, open the selected channel and it will begin to play in iTunes. Now find the "Radio" icon and make sure it matches the name of the designated channel. Dragging and dropping the channel into your radio playlist eliminates the need for knowing what the radio station URL is.

More iTunes Radio Options

While you can get creative and add your own radio channels, there also is a method of expanding the list of default channels. To do this, you need to instruct the program to play radio stations discovered using your Web browser. First, click the "Edit" tab at the top of the screen, then scroll down and select "Preferences." From here, you will notice a series of tabs. Locate "Advanced" and click it.

In this menu, you will navigate to "Use iTunes for Internet music playback." Click the button that reads "Set" to the right side of the text. Your iTunes application has now been configured to automatically play music streams.

Get Even More Radio Stations

Having the ability to stream radio programs is a great benefit for iTunes users. The iTunes default library is impressive and continues to grow. At the same time, these numbers have been restricted due to complications with a few radio stations. Many stations elect to broadcast in the format of Windows Media Player (WMP) or Real Player. Because neither are compatible with the iTunes radio function, this limits the amount of available stations.

Die-hard radio lovers can access more stations through third-party companies. Radio Shark is a device released by the company, Griffin Technology. It is compatible with PCs and Macs and capable of recording all AM and FM broadcasts in real time. This amazing device also allows you to transfer or sync any broadcasts to your iTunes application or iPod.

Podcast, Podcast, and More Podcast

Welcome to the wonderful world of podcasting, the latest great feature of iTunes brought to you by Apple. So what exactly is this new feature? The actual definition of this feature ties heavily to a method of broadcasting through your iPod.

More and more individuals are using podcasts to publish various forms of content on the Internet. This is done with graphics, audio, video, and much more. The technology used for this advanced method of online communication happens through Real Simple Syndication (RSS) feeds. The actual podcasts are composed of XML scripts, a popular language of the Web. These XML files reference a type of media called "enclosures" to "'podcatchers." The podcatchers then translate the XML files and download the linked enclosures.

While there are a few similarities, podcasts are quite different from Internet radio. Radio programs are typically live broadcasts; podcasts have no time restrictions, meaning they can be accessed and played at any time. Your favorite episodes of podcasts can be delivered automatically when making a subscription through the iTunes Store. Podcasts are frequently updated and always available. By using the iTunes software, you will have just as much control over this great feature as you do with the music and other content on your iPod. The best factor of all is that podcasts subscriptions are free via the iTunes Store.

Working with Podcasts

The overall aspects of Podcasts may appear somewhat challenging and foreign to most. The truth is that they can be very entertaining and beneficial for some. Learning to operate your podcasts feature is not actually as difficult as it may seem. Here a few steps to help you set up:

Open your iTunes software and locate the "iTunes Store" tab to the left of the screen.

Click it.

Just above that link, you will notice a tab for Podcasts. Click it. Location of this tab depends on your version of iTunes — these steps relate specifically to iTunes 7.

This will open a menu on the iTunes Store filled with selections New Releases, What's Hot, and Staff Favorites. Staff Favorites consists of featured podcasts that have been chosen by the editors of iTunes. You will find a list of topics to the left: Arts, News & Politics, and Technology. In these categories, you will find subtopics that give you even more options. You also may browse through popular podcasts providers, such as Comedy Central and The New York Times Company.

After browsing through the selection of available podcasts, click on "Get Episode" at the right of the page. The iTunes Store will initiate the download and store it into your "Podcasts directory." As you gather more, your podcasts will be categorized by name and filed by episode. When the download of the podcast is complete, you can begin to play it by double clicking the text file. If you happen to like the program and choose to listen to it regularly, simply click the "Subscribe" tab beside it and you will automatically receive it.

Sharing Music Legally

With the abundant amount of music and other content offered by iTunes, it is natural that you may want to share these perks with your friends, saving them a few dollars. Like everything, sharing has its limits — something you should be aware of when distributing your iTunes content.

Apple has made many adjustments and set many restrictions since the initial release of iTunes. Customers took advantage immediately of the feature that allowed users to share music over the Internet. This brought up a number of copyright issues, and the sharing feature was removed with the introduction of iTunes version 4.0.1. The new edition limited the ability of

Internet sharing to streaming — meaning the songs could be accessed and played, but not duplicated — similar to the way a podcast or online radio show operates.

How Sharing Works

The iTunes sharing feature works when you are signed on to the Internet. It allows you to access files from another computer that has iTunes open as long as you are both on the same network. This feature also permits others to access your library when your iTunes application is open. iTunes sharing is beneficial for many apparent reasons, considering all the content available from the online store. You are liable to run across a song you have not heard in years thanks to the sharing feature.

Versions of iTunes 4.0.1 and later come equipped with the sharing feature. All that is required of you is to have an Internet connection, be it dial-up or through a high-speed modem.

To see how it works, follow these instructions:

♪ The first step is to enable the sharing feature. Find and select "Edit" at the top of your display screen.

♪ In the edit menu, select "Preferences."

♪ Scroll through this menu and click on the "Sharing" icon.

Your sharing options will now be listed. The first option you find instructs iTunes to search for shared music. By checking this box, iTunes will scan your local network in search of users who have enabled the next option, "Share My Music." This allows the music in your library to be accessed by others. Of course, these other computers must have iTunes open for this to work on their end.

NOTE: Firewall and other anti-virus applications may block the sharing feature. If you have this type of program installed, be sure to check the configuration settings to learn whether or not iTunes sharing is permitted.

iTunes allows you to control your limitations of sharing as well. You can decide who gets free access and how much they get. You can allow them access to your entire library or designated playlists. If you are willing to give access only to playlists, you can customize the certain amount of shared songs. Playlists of your favorite podcasts and radio shows also can be shared over a network.

Listen to Someone Else's Music

Now that you have learned how to share your music, it is time for you to reap the benefits of this feature. As iTunes goes on a search for music, it may find a single library or several. When shared music has been located, it will display in the panel to the left of the iTunes screen. Click the entry or entries to find exactly what libraries are available. Once you have found a library of your liking, click it and iTunes will load the music.

Once the shared library has been accessed, it will operate just like the one you own. Simply highlight a song or entire playlist and click the "Play" button or double click it.

NOTE: Sharing music is a form of streaming. Someone else's playlist can never become a part of your own library permanently. Ratings cannot be placed on shared music. To end a shared playlist, simply click the "Eject" button beside it. All shared activities will be automatically halted when you close the iTunes application.

Sharing Restrictions

The iTunes sharing feature is great for music lovers, but there are a few limitations of which you need to be aware. Neither you nor another individual may access or play files that are protected — this includes any content that may have been purchased from the iTunes Store. The exception here is access for other computers in your home. You will have to manually grant authorization by referring to your account with the iTunes Store.

While you are permitted to share audiobooks purchased from the iTunes Store, those bought from Audible.com may not be shared. Music from a shared library cannot be included into the cycle of your party shuffle.

Apple has legally attained rights to all songs available on the iTunes Store. This has been made official through negotiations with the five major record labels that have offered their catalogs. This is certainly a safe option for downloading music, as opposed to free music sharing software such as LimeWire and Napster, in which you are at the extreme risk of copyright violations.

Individuals who purchase and download music have rights of their own concerning what can be done with the content. Users with an account to the iTunes Store have the permission to transfer and play songs on up to five different computers. Each computer is then allowed to burn content onto a CD at an unlimited rate. This has proven to be a sufficient regulation for most users of the iTunes Store. On the other hand, users may create an unlimited amount of different accounts to the iTunes Store on any computer.

Case Study: Suzanne Lieurance

Suzanne Lieurance

Author

The Working Writer's Coach

I have been writing all my life. I always knew I was meant to be a writer even when I was doing other things (like teaching) to make a living.

I am primarily a children's writer (see my books at **www.suzannelieurance.com**). I have written over a dozen published books for children. My latest children's book will be released in May (2008). It is called **The Locket: Surviving the Triangle Shirtwaist Fire**. It is part of Enslow Publishers' new Historical Fiction Adventures series.

In addition to writing, I am a writing instructor and writing coach. I am also the founder and director of the National Writing for Children Center, an online resource for children's writers, parents, teachers, kids, and just anyone interested in children's publishing.

As The Working Writer's Coach (**www.workingwriterscoach.com**), I help people who live to write become "working" freelance writers and make a living doing what they like to do best — write! I also help business owners become noted experts in their field by writing a book about their business.

I have been interviewed for Writers in the Sky podcasts and I have also been a guest host for their podcasts. But most of my experience with podcasting has been with my own podcasts. I have a podcast at the National Writing for Children Center site (**www.writingforchildrencenter.com**), called "Book Bites for Kids." This podcast is also a LIVE radio show on **blogtalkradio.com**. I have also been an instructor for the University of Masters (**www.universityofmasters.com**), where I offered a monthly podcast about freelance writing. They are great for promoting my books and the books of other authors.

They are also a great way to teach or coach. Students or coaching clients can simply download a recorded podcast and save it to their MP3 player and listen to it whenever they want.

Aside from podcasts, I find iPods attractive because they are small, so you can take them with you anywhere and listen to recorded conferences, classes, and workshops, or music. So, they are not just a tool for enjoyment — they are a learning tool!

Case Study: Suzanne Lieurance

I like listening to things recorded on my iPod when I am riding in the car. I am not the only one. I think people call this the "Auto University." I also listen to it at the gym when I am walking on the treadmill.

I do not use my iPod for music, yet it does everything I want it to do. However, newer models keep coming out all the time and they can do more and more, and prices are incredibly low for all the features you get. I do not know all features and capability of the newer models.

I am not familiar with any other brand of MP3 player, but I am an Apple lover for computers, too, so I tend to go with anything Apple.

Working with Photos
and Video

We all recognize an MP3 player as a device capable of cranking out a large amount of music, whether it is an iPod or another brand. Where the others stopped is where Apple continues to run and deliver amazing features. One of them happens to be photo capability that seems to get more advanced with every iPod.

Apple became picture perfect with the introduction of iPod Photo. Not much larger than your average deck of poker cards, this spectacular device still delivered the goods and continued the iPod craze. The device features a crystal clear LCD display screen, giving you 65,536 colors to view your pictures in. The easy-to-use backlight function allows you to flip through the photo gallery, regardless of your environment.

The classic click wheel on this iPod makes for easy navigation. The new display comes in crisp color and brings your photos to life. Your iPod will resemble a real-life photo album, as your new screen can display up to 25 full-sized thumbnail pictures at once.

Pictures are its specialty and the iPod Photo is up for the task. This device is sure to boost your multimedia experience to the next level. Make use of the complimentary AV cable and attach it to your television. You can view pictures or amaze your loved ones with breathtaking slideshows.

Transferring your favorite pictures is simple with the iPod Photo. Files from your "My Pictures" folder are imported directly to the iTunes library. Your previously organized photos remain intact and can be easily accessed in original order and format on your iPod.

iPods with photo features eliminate the need for those worn out pictures you have been carrying around in your wallet. Whip out your handy iPod and show them off in crystal clear color, as they were meant to be viewed. Your picture gallery can also be easily accessed from your favorite photo editing application, such as Photoshop or Corel Photo Center. Simply instruct the iPod where to extract the pictures from, and the device will do the rest.

What It Takes

Before getting all psyched to put pictures on your iPod, there are a few things to be aware of. Importing photos to the device requires having and following a few key elements.

System and Photo Accessories

First and foremost, you will need a Mac or PC, preferably with a color monitor. Of course, the iPod Photo or any later model with photo capability is required. You will also need a photo storage application installed on the computer. Programs such as Adobe Photoshop or Ulead Photo Impact will work. If you do not have any of this software, the default folders on your system will operate the same. Photos can be transferred from a "My Music" folder (PC users) or an "iPhoto" folder (Mac users).

Properly Formatted Files

Your photo files must be saved in the correct format. In most instances, typical formats for digital cameras and Web pages are compatible with the iTunes software, yet this also depends on the operating system you are running. Common photo image formats in Windows are PNG, JPG, TIF, SGI, GIF, PSD, and BMP. Those on a Macintosh are JPG, GIF, PICT, BMG, PNG, JPG2000, and PSD. All of these formats should be compatible with both the iTunes software and your iPod.

Upload the Pictures

After making sure you, your computer, and your software meet all the requirements, you are one step closer. Your digital pictures are all in the proper format, and you are ready to get started. It is time move the pictures from iTunes and sync them to your iPod. A good idea would be to configure the device to make copies of the photos for safety measures. Here a few simple steps:

First, attach the iPod to your computer via the USB cable. The iPod will be displayed in the left column of your iTunes program. Click the icon that represents it. A preference menu will display in the program. Find and click the "Photos" tab.

In this menu, you will see "Sync photos from." Enable this option by clicking the checkbox beside it. In this screen you will also select the photo software or folder in which your pictures will be extracted. You have the option to copy every photo in the folder or just select albums and pictures. Click the "Apply" tab after deciding.

You may not want to select any programs found in the "Sync photos from" display and decide to copy photo folders located on the hard drive. To do this, hit the "Choose folder" tab, and then find the appropriate file that contains your photos. You can either sync the entire folder or the files found within it.

If you desire to have every one of your images synced to the iPod, select the "All photos and albums" tab. If you only want certain pictures imported to the device, choose "Selected albums" to choose your favorite picture collections.

NOTE: Every time you perform a sync of your photos, those selected will be upload, and any pictures you have recently added will be detected as well. In the midst of this process, the iTunes program will display a message like "Optimizing photos." Do not panic after seeing this message. This just means that your iTunes software is making adjustments to your photos for the best viewing results. It will create different versions of those photos and optimize them for viewing on the screen of your iPod or television. Before being synced to the iPod, the photos are copied, then stored on then hard drive of your computer.

Full Size Photos on Your iPod

As your photos are being optimized by iTunes for the sync, the selected images are streamlined for a quicker transmission. This method is opposed to larger, high resolution files being duplicated. But if you want those full-sized photos on your iPod, follow these directions:

Connect the iPod to your computer via the USB cable.

Select the iPod icon when it appears on the left side of the iTunes display.

Located in the settings page for the iPod, click the "Summary" tab.

From this menu mark the checkbox beside "Enable" disk use.

NOTE: For this to work, your iPod must be configured to act as a hard drive. The previous step was a brief summary of how to enable it. More detailed instructions will be discussed later in the book.

Now you will find and click the "Photo" tab located on the left side of the iTunes display screen.

You will then notice the text "Include full-resolution" photos. Mark the checkbox beside it to enable this option.

Once the sync is complete, you will have full-sized, high resolution copies of your photos in the hard drive of your iPod. Within the photos menu of your iPod is a subfolder that contains all of the thumbnail images that have been optimized by iTunes. If you wish to strictly work with full-sized photos, these may be easily avoided. Making use of full-sized, high resolution photos is a great idea for photographers and any individual who may need to view pictures while on the go.

Viewing Photos on Your iPod

After finally releasing your photos from your computer, it is time to view the results. Show off your favorite pictures and your handy new iPod at the same time. Looking at the pictures you placed on the iPod is actually easier than getting them on the device. Here is how to do it:

On your iPod, select the "Photos" option found in the main menu. If you decided to sync certain albums, simply maneuver the click wheel to the name of that photo album and press the "Select" button.

A new screen will display with miniature versions of your photos in the selected album. Move the click wheel up or down and highlight the photo you would like to view. If you happen to have hundreds of thumbnails to scroll through, clicking the "Next" or "Previous" button on the iPod will help you move faster through the selections.

Additional Viewing Tips

By getting a little experience, you will quickly learn that working your iPod is fairly simple. The device gives a few different options to perform the same function, making it easy to manage your photos.

Another way to look at pictures is to navigate to the desired photo, highlight it, and tap the button in the center of your iPod. This will actually display a larger version of the selected photo. When you want to go back the smaller sized picture, simply tap the "Menu" button. Sifting through your pictures can be done by tapping either the "Next" and "Previous" buttons or by rotating the click wheel clockwise our counterclockwise.

Creating and Playing Slideshows

The iPod Photo brought forth many additional features to Apple's mega multimedia player. Not only can you import and view photos on the device, you can also customize your albums. One of the most notable upgrades is the ability to run slideshows of your favorite photos. A slideshow is simply a view of photos in motion. The pictures slide in a configured sequence from one photo to the next. This function is similar to moving photos you may see on someone's MySpace profile page.

Slideshows are a fun way to display your pictures. Before setting up a slideshow, you must make a few adjustments, such as configuring a running time and background music. To begin, select the "Photos" option from the main menu of your iPod.

In the new menu, select "Slideshow settings." This will display several other options that allow you to customize the slideshow to your preference.

To set the duration of the slideshow, select the "Time Per Slide" menu. This will instruct the iPod on how long to display a picture before moving along to the next one. You have the option of choosing 2 to 20 seconds. You may also elect to navigate through the slideshow manually with the simple tap of the click wheel.

To set your background music, select "Music" from the main menu screen. You can choose to use a song or playlist already on the iPod or create a new one and assign it to the slideshow.

Similar to music on your iPod, the photos in your slideshow can be placed on repeat or the viewing order may be shuffled. Special effects can also be added to make your slideshow much more spectacular. To do this, select "Photos" from the main menu and then choose "Slideshow settings" in the following screen. From there, select "Transitions." In this menu, you will find an array of effects that will be applied to the slideshow when photos change from one to another.

Models such as the iPod Photo allow you to view your slideshows on the device as well as on a television set. To be certain that the slideshow will display on the screen of your iPod, make sure it the iPod is configured to do so. Do this by turning off the "TV OUT" option on the screen in the "Slideshow settings" menu. You can also configure the iPod to ask if you want the slideshow to be displayed on the device or on a television before each run.

When you have finally set the slideshow options to your liking, simply select the picture or album you wish to start with and press the "Play/ Pause" button, which will initiate the show.

You can also freeze the slideshow by pressing the "Play/Pause" button. If the slideshow is not moving fast enough, you can reconfigure the duration, or simply move it along manually by tapping "Next" or "Previous."

Experience the Slideshow on Your TV

This option requires the appropriate adapters that connect the iPod to your television set, VCR, or DVD player. When the connections are in place, there are a few more configurations that need to be made:

From your iPod, choose "Photos" from the main menu.

In that menu select "Slideshow settings."

In the following menu, you will choose "TV," and make sure that it is set to "On." This will instruct the iPod to output the actual slideshow on a TV.

Now select the broadcasting standards of your locality. For example, if you reside in either North America or Japan, select "NTSC" for your "TV Signal." If you reside Australia or Europe, you would set the "TV Signal" to "PAL." If the signal for your location is not listed, you may want to check the manual of your television to learn its preference.

Next, power on your television set and choose the iPod's video input source. This is done in the same manner in which you would instruct the TV to recognize signals from your VCR or DVD player. In most cases, you would tap the "Input" button on the television or remote control to switch the current signal to the video source of the new device. On a TV with cable connected, this could be as simple as turning it on "Channel 3."

Following the directions, setup the slideshow on your iPod and tap the "Play/Pause" button to initiate it. Your vibrant photos will then run on the television with the matching theme music you applied to the slideshow.

Your iPod will go far and beyond when it comes to managing photos, but there are a few things it is unable to do. Your iPod cannot pull photo images directly from a CD. Any files you take from a CD must first be imported into the computer's hard drive.

As we learned earlier, your iPod connects itself with a computer during the syncing process. Setting your iPod to manual mode is a good trick for music, but not as effective when it comes to photos. Your photos can only be synced from one single computer. Importing pictures from another computer is a gamble you should not risk, as the new pictures will replace whatever you currently have stored on the iPod.

Photo Meets Video

The release of the iPod Video was met with great anticipation. It quickly trailed the iPod Photo, and there was much speculation concerning the capability of Apple's latest super device. What type of video would it support? Where would they come from? The iPod Video simply revealed notable upgrades to the previous model and introduced a video outlet to stand aside the globe's hottest online music store.

New and Improved Feature

The iPod Video quickly became one of the best selling of all iPod models. The luxuries of music and audiobooks had been combined with bigger and better photo capability. This iPod comes with a storage gallery for photos that easily holds full-sized thumbnails. The slideshow feature is another great addition that separates this model from the iPod Photo. The biggest change is video playback. You are now able to watch the latest music videos, and even full length feature movies. The battery life on this model has been extended to 20 hours, a great attribute for those looking to get the most out of their iPods.

The iPod Video also evolved with the size of its color screen, which displays brighter photos with sharper text. This model is available in the colors of black and white, and is slightly bigger than the previous model. It comes equipped with many new applications, such as one for the Screen Lock function. Storage space makes this iPod very attractive, with a 30 gigabyte and 60 gigabyte version, which you will typically find at the same price.

While this model was made to introduce video to the portable multimedia industry, the iPod Video has many new photo capabilities that make it a winner as well. Similar to earlier models, like the iPod Nano and iPod Photo, it is able to import digital images from a computer, camera, or card reader. A great way to do this is by getting aid from a good program, such as Camera Connector by Apple.

Importing and saving photos is a bit different when comparing the iPod Video and iPod Photo. Capacity is the major difference. The iPod Video is large enough to hold up to 15,000 songs, 25,000 pictures, and as much 150 hours of video.

Videos at the iTunes Store

When in dire of need of movies for that dreadful plane ride, the iTunes Store is your one-stop shop for the latest releases. You can pick a movie that was just released on DVD or go with one still playing at the theater. Choose from hundreds of films, thousands of music videos from today's hottest artists, and tons of your favorite TV sitcoms.

Surfing through the endless sea of videos on the iTunes Store is easy. New releases are displayed on the home page, along with a list of best selling videos. Apple gives a taste of what is ahead with detailed film descriptions and preview capability before making a purchase. There is also a "Power

Search" feature that enables you to look up movies by title, director, actor, year of release, and rating.

Purchasing and Downloading Video

The iTunes Store is the ultimate super center of virtual reality. You will be amazed by the extensive catalog of music, movies, and much more. The process of getting video from the iTunes Store is nearly identical to music. Let us review the steps:

First, open the iTunes software on your computer.

You can access the "Store" at the top of left side of the iTunes screen.

Browse the selection of videos from the front page of the store: Movies, Music, and TV Show. If you are looking for something in particular, type it into the search box in the top right corner.

After selecting the video you want, click on the image.

Click the "Add Video" or "Buy Now" tab to initiate the purchasing process.

Once you enter you iTunes Store account information, the download will commence.

Syncing Video to Your iPod

After purchasing and downloading movies from the iTunes Store, you will want to make sure they work. Video content purchase from the iTunes Store cannot be viewed on the iTunes program. This can only be done on your iPod or default video player, which would be QuickTime or Windows Media Player for most users.

You can sync movies, music videos, and TV shows to your iPod, either automatically or manually. Follow these steps:

Connect your iPod to the computer via the USB cable.

Open the iTunes program.

Click the "iPod" icon when it displays to the left of the iTunes screen.

Then select the tab that applies to your video content: Movies, TV Shows, or Podcasts.

You have the option of syncing all videos in iTunes, those in certain playlists, or select videos according to content.

After making a choice, click "Apply" to begin the sync.

Observe the "Do Not Disconnect" message as the content it being synced.

When the process is complete, click the "Eject" tab in the "iPod" menu.

You can now disconnect the device.

Manual Video Sync

In some cases, you may wish to manually sync your video content. This will only import the items you select, as opposed to everything in the iTunes library. Follow these steps to take more control of your video:

Connect your iPod to the computer via the USB cable.

Open the iTunes program.

Click the "iPod" icon when it displays to the left of the iTunes screen.

From the iPod menu, click the "Summary" tab. Be sure to mark the "Manually manage music and videos option."

Next, click the "Apply" tab.

You can now manually drag and drop video content into your iPod.

How to Play Video on Your iPod

Now that you have a decent collection of movies and music videos on your iPod, let us cover the steps how to view it on the iPod:

From the "Main Menu" of your iPod, scroll down and select "Videos."

In the next screen, you will find all the options for video content: Video Playlists, Movies, Music Videos, Video Podcasts, and Video Settings. In Video Settings, there will be options for "TV Out," which allows you to display iPod videos on a TV; "TV Signal," which can be set to PAL or NTSC; and "Widescreen," which can be toggled on or off.

For now, we will highlight and select "Movies."

The screen will display a list of all the movies you have downloaded from the iTunes Store and synced to your iPod. Highlight a movie, press the "Select" button, and it will begin to play.

iTunes Videos on Your TV

Similar to the slideshow experience, videos and the iTunes software can be synced to the iPod and then displayed on your TV. This requires a video cable that outputs the content. This cable is easily connected from the headphone jack of the iPod to the appropriate ports of your television, VCR, or DVD player.

If watching videos from iTunes becomes a habit, there is a way to get better quality. Going with video files in MPEG-4 format would be the best bet. This format supports a much larger size. The estimated limit of pixels is around 230,400, more than enough to display crystal clear, widescreen movies on your television set. Try MPEG-4 against another format and you will certainly observe the difference those additional pixels make.

Video files on the iTunes software on their way to your iPod may warrant the issue of compatibility. You may run across a few video files downloaded from the Web or other sharing forums that are not compatible with your iPod. The best method for figuring whether or not certain video is accessible by the iPod is to drag and drop the selected file into the library of your iPod. All systems are a go if the video file is able to be copied and transferred to the iPod.

NOTE: A simpler way to check for compatibility is to add the video files to your iTunes library. Highlight the file in question, right click it, and choose "Convert Selection to iPod." The iTunes application will then notify you of whether the video file is compatible. In the event that the file is not compatible, the program will convert it into a format that works for you.

NOTE: File conversions stand the chance of decreasing video quality. Do your best to download video files that are already compatible with both the iTunes program and your iPod.

Home Videos on Your iPod

You will certainly enjoy the large amount of video content available from the iTunes Store, but what about your home movies? Do not worry; with a little know how, those too can be transferred to an iPod and stored right in your pocket.

Home Videos for Windows

The default video handling on Windows would be Media Player. The default program for view editing is Movie Maker, the application we will cover in regard to your home movies. The traditional setup of Movie Maker is not intended for iPods. This is because it saves movie files in either a DV-AVI or WMV format. To extract your home movies from this application, you must first convert them into a file format compatible with the iPod.

You can begin by saving the movie file to your computer in the usual manner, preferably in a high quality WMV format. The aid of a third-party application is required to make for a successful file conversion. I suggest using a tool by the name of WinFF. It is free to use and downloads quickly to your computer. From here, all you have to do is follow these simple steps:

Once the move has been saved, open the WinFF application from your desktop. Add WinFF to the list of movies you want converted to your iPod (you can convert movies one by one or as a group).

In the "Convert to" menu, select "Convert to Xvid for iPod."

Now browse your computer's iTunes Movies folder in the "Output Folder" menu of WinFF.

Finally, click the "Convert" tab to begin the process.

By saving your home videos in the iTunes Movies folder, you automatically enable the process of accessing files from WinFF. This means that your iPod will always get the home movies since the files are stored in iTunes.

When you have completed the final step, open the iTunes software and make sure the home movie is stored in your library. If so, it can then be synced to the iPod with the rest of your video content.

Home Videos on a Mac

Home videos on your Mac computer can be viewed on your iPod as well. This requires the use of QuickTime version 7 Pro or later. If this application was not included with your iTunes, you can visit the Apple Web site for a free download. From there, you will follow these directions:

First, open QuickTime version 7 pro.

Click the "File" tab at the top of the screen.

In the File menu, highlight the movie you would like to view on your iPod.

Double click the selection of hit "Open."

From the "Export" drop down menu, select "Movie to iPod" – 320 by 240 resolution.

Next, click the "Save" tab.

You will notice a progress bar indicating how much of the process has been completed. The longer the movie, the larger a file it will create. This translates to a longer exporting process.

When the process is done, Quick Time will display a new version of the home movie in the form of an icon. Double click the file to begin the movie and inspect it for flaws. If everything looks good, then it is time to transfer the home movie to your iPod. Follow these steps:

Open the iTunes program.

Import the home movie from the appropriate folder on your desktop or manually drag it.

Now, connect your iPod to the computer.

Click the "iPod" icon when it displays to the left of the iTunes screen.

From here, you can follow the previous steps on video sync to transfer your data automatically or manually.

Case Study: Donna Kshir

Donna Kshir

Freelance Writer — Author

For more about Donna Kshir and her writing visit:

MySpace Page:

www.myspace.com/donnakshir

FanSite:

http://dkshir.tripod.com

She Unlimited magazine

http://www.sheunlimited.com/2007/10/31/meet-us-donna-kshir-contributor/

Writing has always been a big part of my life, but it has only been the last three years that I have been taking it seriously and began freelancing for several online and print magazine publications. Recently, I teamed up with Barbara Mastriania and Lauren Lawson to become a contributing writer and columnist for my hometown newspaper, *The Record*.

My upcoming novel, *My Life's Fight*, is based on pro MMA fighter Mark Bailey's life. Mark learned at a very early age that losing was not an option, his father would not allow it. He was taught to never show emotions and that winning was everything. As a child, Mark was afraid to be a failure as he feared his father. Throughout Mark's life, he has carried extra baggage from his tough childhood. After a short stay in prison, Mark found God and it changed his life forever.

Case Study: Donna Kshir

In addition to my book with Mark, I have my own upcoming book series called, "The Dear Diary Series." The first book in the series to be released is *Dear Diary-Oh Mother. Dear Diary-Oh Mother* is about an elderly woman, Jade, who looks back on her life and relives the abuse and pain caused by her stepmother Alice. In the end, Jade must forgive her stepmother or live with hate in her heart forever. Both books are scheduled to be released in Spring 2008.

I began contributing to *She-Unlimited* magazine in July 2007, as an entertainment writer. I cover hot topics surfacing in the news, as well as conducting music, movie and book reviews, and interviews.

Being a writer is a lot of hard work. The biggest problem is getting your name out there and, more importantly, get noticed. Knowing how hard it was for me and the struggles and sacrifices I made, I wanted to make the process easier for others. I recently opened my own production company, donna kshir productions. My intent to help other writers live their dream and be published. The most important part of my life is being a wife and mother. I have a wonderful husband Roger and together we have two wonderful children, a son, Roger, and a daughter, Ashley.

My life is very busy and I am always on the go. Two years ago, knowing my busy schedule and my love of music, my kids bought me an iPod Nano. I instantly loved it. It was small, light and convenient, but most importantly I loved the fact of only downloading the music I wanted to hear.

To date, I own several iPods. I use my iPod Shuffle the most. I walk five times a week for an hour a day. My iPod Shuffle is small and light. I clip it to my shirt and go. I do not have to worry about carrying it or putting it in my purse or pocket. It can be used anywhere: at home, exercising, relaxing, and even in the car. I recently bought an iPod adapter for in my car. I no longer have to store or carry around bulky CDs. I just plug the adapter into my iPod and listen to my favorite music through my car radio.

Currently, I own four iPods. I personally have not had any problems with them. I love the variety of colors, shapes, sizes, and styles. iPod makers have already outdone themselves, as they added the iPod Touch. I feel there is no room for improvement.

I love my iPod Touch, but my favorite model is my iPod Shuffle. There are times I do not carry a purse or bag and I need my hands to remain free. With the iPod Shuffle you do not need to worry about carrying it around, you just clip it to your clothing and go. I love that feature.

The model I would recommend for someone looking to buy their first iPod would be the Nano. It is inexpensive and I feel it is the easiest system to learn, especially for small and young children or first-time users.

Hands-On Training

By now, you have learned that the iPod is much more than your typical MP3 player. The ability to playback a large amount of audio tracks was its original claim to fame, but it has since moved on to bigger and better features. Videos, podcasts, and audiobooks are just a few of the upgrades added to make this device even more extraordinary.

It is no surprise that some people find certain features of the iPod more favorable than others. There are music lovers who went out and purchased one just to take their party on the road, and there are movie fanatics who need to watch their favorite films and TV shows anywhere, at anytime. The evolution of podcasts has seen more business-minded individuals using the device to flourish their online ventures. Of course, there are some who are multi-functional with their iPod and regularly use every single feature.

The time frame for totally mastering an iPod depends on the individual. You may choose to take it slow, making yourself absolutely familiar with importing, burning, and playing music at first. Then you may choose to move on to other features, such as video and podcasts. Having come this far in the handbook, you are well on your way to conquering every single detail and knowing every feature of your iPod.

Learning the basic controls of your iPod is a must. Familiarizing yourself with all of the features of the iTunes program and iTunes store is also a plus if you are looking to get the most out of your device. Since we have covered the key elements of getting started and gone in depth with handling iTunes, it is time for the next level — hands on practice with your iPod. This chapter will help you apply everything you have learned previously and make your portable multimedia experience even more incredible.

Playing Music on Your iPod

Regardless of what you have planned for your iPod, music is a great place to start. Everyone has a song they love, be it something new or an old-time classic. Besides, playing music on your iPod is very easy. Learning this step makes everything else a bit simpler, as many other functions are quite similar to how you play music on an iPod.

Performing any action on your iPod comes with use of the of the handy click wheel. Gently moving your thumb over the wheel is certainly the easiest way to navigate through the menu screens. Browse these menus for the song you wish to hear and press "Select" (the button in the center of the iPod) after finding it. If you change your mind and choose to play another song, simply press the "Menu" button on the click wheel to move back to the previous screen. Once you are set on a song, album, or playlist, hit the "Play/Pause" button on the iPod and the item will instantly start up.

Playing Music During a Charge

While the iPod is accepting a charge, it may not play music unless configured to do so. This is the case if you are charging the iPod with a USB cable that is connected to a computer. When plugging the device to the USB adapter, you will see a battery icon indicating the charging process along with this message: "Do Not Disconnect."

You can press the buttons on your iPod all you want but nothing will happen; the device will continue to charge and ignore your demands until the process is complete.

By making the proper adjustments, your iPod will be able to play music via the computer as it accepts a charge. Here is what you need to do:

Step 1: First, connect the device to your computer with the USB cable.

Step 2: Next, open the iTunes application.

Step 3: Click on the "iPod" icon that displays on the left side of the screen.

Step 4: In the "iPod Summary" menu, navigate your way to the "Manually Manage Music" option and mark the checkbox beside it.

Step 5: Next click "Apply."

By following these steps, you can control your iPod and all its contents through the computer. Even if you have made adjustments to the iTunes library, the library on your device remains intact and allows you play every song, music video, or movie you previously synced to it.

NOTE: Using a separate, plug-in power adapter gives you more freedom during the charging process. This allows you to operate your iPod normally as if the charger is not even attached. This could be great option for someone not looking forward to making adjustments within the iPod.

Optimize Your Playing Time

The battery life of most iPods is exceptional. There are also a few things you can do to make sure it exceeds the norm and get the most out of your playing time:

1. Instead of waiting for the iPod to shut itself down when not in use, turn it off by pressing the "Play/Pause" button in the center of the click wheel.

2. Place your iPod on "Hold" by enabling the switch.

3. Minimize the use of your backlight.

4. Disable the Backlight Timer on your iPod. To do this, select the "Settings" menu, "Backlight Timer," and make sure it is off.

5. Turn down the brightness of your display screen. Choose the "Settings" menu, select "Brightness," and maneuver the wheel toward the left to reduce the color.

6. Only press "Next" or "Previous" on the iPod when it is necessary.

Remembering to manually turn off your iPod and cautiously flip through your playlists may be a bit difficult for some. These things are not required to get longevity and enjoyment from your iPod. Equipping yourself with a few good tips will not hurt anything, either. If anything, taking note of these factors can at least save a few bucks on your battery.

Podcasting with Your iPod

In Chapter 5, we discussed podcasts and how they are quickly becoming one of the iPods most widely used features for various reasons. Cable companies across the nation offer movies when you want them — Apple rivals that with podcasts on demand (your favorite radio shows and Internet programming in the palm of your hand).

Unlike online radio, where you stream in to live shows and run the risk of missing important segments, podcasts can be listened to at any time, from beginning to end, giving you the power to pause, fast forward, or rewind at will.

What has made podcasts so popular is the versatility they offer. The shows may come in the form of headlining news topics, music reviews, online talks, or audio blogs from ordinary people. The iTunes Store gives you many different podcasts from world renowned broadcasting companies such as CNN, The Wall Street Journal, and ESPN. You will also find shows from popular celebrity personalities like movie critics Ebert & Roeper, Adam Curry, and a few lesser-known individuals who have gained a bit of notoriety by way of the podcasting trend.

How to Get and Play Podcasts

Since we have already covered how to access free podcasts from the iTunes Store, let us now discuss the procedures of getting them on your iPod. We will begin with a brief recap of finding your podcasts on the iTunes Store:

1. Open your iTunes application and click on the "iTunes Store." Browse the selection of podcasts and click the one you would like to view. The broadcast will be downloaded to your iTunes library and then you can subscribe to it.

2. Now, attach your iPod to the computer via the USB cable.

3. Click the "iPod" icon in the iTunes screen.

4. From the next menu, click the "Podcasts" tab.

5. To automatically import podcasts, choose the "Sync" option.

6. Select the podcasts you want synced to the iPod. Choose all of them or only certain broadcasts.

7. Click "Apply."

8. Find and click "File" at the top left of your iTunes screen.

9. Scroll down this menu and select "Sync iPod."

10. Observe the "Do Not Disconnect" message. When the message goes away, the process is complete and you may then disconnect.

11. Finally, click the "Eject" tab next the iPod icon.

12. Navigate from the main menu of your iPod and find "Video."

13. Next, select "Video Podcasts." Find the broadcasts you would like to view and press "Play."

Creating Your Own Podcasts

For some, podcasts are a great vehicle toward the broadcasting world and beyond. Others use them to enhance their blogging experience or express opinions in a more creative light.

You could have a loyal fanbase with faithful subscriptions to your podcasts. All it takes to get started is a Mac or a PC, capable recording software, a Web server to host the broadcasts, and your motivation and talent.

The podcasting feature of the iPod has been useful for students and professionals. Record important lectures from your college course or capture exclusive interviews to push your broadcasting career.

More on Podcasting Tools

To record broadcasts while out on the road, you will also need a microphone. While several types may be compatible with your iPod or computer, all will not be of quality, so you should be selective. The microphone included with your computer might have been appropriate for play time with the kids, but is not recommended for someone looking to make an impact in the way of podcasts.

Of course, you want a microphone that works well, but there is no need to go overboard on the budget. Headsets with USB connectors are effective and affordable. These devices are lightweight, professionally stylish, and designed to filter out background sounds. We will go into further detail on recording devices and how they work in the chapter on iPod accessories.

Podcasting Software

Another important element of getting your broadcasts on other people's iPods is the use of podcasting software. These types of programs work hand-in-hand with a microphone and make recording your content easy, whether you are recording from home or creating out in the field.

One notable recording program to use is TuneStudio, designed by the company Belkin. This software acts as a fully functioning workstation for recording and editing your podcasts. It uses the iPod as the storage unit and enables you to work directly on the device.

This program has gained popularity and fame by way of its mixing feature. TuneStudio is actually a four channel mixer that permits you to attach microphones or recording devices.

Every channel within it is equipped with a three-band equalizer with its one stereo control and recording level. You can also compress audio files and monitor the status of audio and recording with a switch found on the mixer. The podcasts you edit from the program are then sent directly to your iPod in a high quality, 44kHz WAV file format.

Belkin's TuneStudio can be purchased for about $180 online. This program comes equipped with the Ableton Live Lite digital audio application and is compatible with a Mac or PC computer.

NOTE: The TuneStudio will only work with the iPod Video. A worthy alternative to TuneStudio is a program by the name Audacity. It works well at creating podcasts from the stages of recording through the final editing. This program adds pizzazz to your shows by allowing you to combine them with catchy introductions and theme songs. Audacity has become popular with podcaster because it is an open-source program, meaning it is free to use.

The interface of the Audacity program is easy to operate. All the essential functions are displayed on one screen, making the editing process a breeze. This program lets you remove mishaps or other recording errors from your podcasts. Audacity is great at eliminating those annoying background sounds and any static that may show up in the recording.

When you have finished recording and editing your podcasts, Audacity lets you export your creation in a variety of formats such WAV, AIFF, and MP3. Podcasts exported in MP3 format require an MP3 encoder to aid in the process, while WAV and AIFF files are automatically compressed by the iTunes software.

Unleash Your Podcasts

After coming up with a clever concept, transforming that into a podcast, and making all the necessary adjustments, it is time to share it with the world. This book previously mentioned the relation of podcasts and RSS feeds. It is necessary that your Web hosting server support this feature for people to view your podcasts. After getting the URL for the feed, visit the iTunes Store and select the "Podcasts" tab to the left of the screen. On the "Podcasts" page, find the "Publish Podcast" tab on the left hand side and click it. From there, go into the URL of the podcast host and select "Continue."

NOTE: After hitting "Continue," you will be prompted to enter a user ID and password for the iTunes Store. This is a requirement for publishing your own podcasts, although you can browse, subscribe, and download them on iTunes without an account to the store.

If you have an account with the iTunes Store, you are halfway there. Apple will then review your podcast request and promptly add it if it is in agreement with their terms and conditions. From there, you wait for people to subscribe and watch your creative broadcasts on their iPods.

Podcasts on Your iPod Shuffle

There has been some debate concerning the podcasting capability of an iPod shuffle. The issues come from attempting to load podcasts from another computer that are in MP3 format. By doing this, all other MP3 files on the iPod may be deleted and replaced by the new podcasts.

While there is software to combat this problem, the best method is to play podcasts that are in MP3 and WAV formats, opposed to AAC or M4B. The iPod shuffle will not automatically sync podcasts, so you will need to add them manually by dragging and dropping them into the device like any other item.

NOTE: You can play and pause podcasts on an iPod Shuffle like any other mode. After pausing and press play, your podcast can take up to six seconds to resume.

Playing Audiobooks on Your iPod

Many of us have acquired a love for reading. Several more will admit that while they enjoy a good book they simply do not have enough time to set down and flip through it. Your iPod puts enjoyment back into the experience with the audiobook feature. You can now listen to intriguing tales of your favorite authors during a jog, driving in the car, or even lounging around the house.

A variety of audiobooks are available from both the iTunes Store and Audible.com. These files come in many different formats, but typically integrate better with an iPod when in MP3 or WAV.

Once audiobooks are synced to an iPod, they can be found in one of two locations. If the files are the MB4 format, they can be accessed through the "Music" menu and will appear in "AudioBook." This will automatically be the case even if you did not label the genre as audiobooks from iTunes. If the audiobooks are in MP3 format, you may have to search through the iPod to find them.

Audio files can be played just like music on your iPod. Files in MB4 format give you extra options and allow you to make adjustments to the speed. The speed can also be adjusted as you listen by maneuvering your thumb over the click wheel.

One neat thing about audiobooks from Audible.com is the bookmark feature that helps you keep your place.

NOTE: Bookmarks are compatible with audiobooks that have AAC file extensions and those purchased from the iTunes Store with M4B file extensions. Bookmarks are not available for audiobooks in MP3 format. If any of your audiobooks are in AAC format, you can make them bookmark compatible by converting them to M4B.

How to Bookmark an Audiobook

Setting a bookmark for an audible file is as simple as placing your thumb in the center of the device. When you are playing the file, simply press "Stop" or "Pause" on the iPod. When you start the audiobook up again, it will automatically begin to play from the last stop point.

Audiobook Secrets with MP3

MP3 files and audiobooks prevent you from creating bookmarks, as mentioned above. This is because your iPod typically recognizes the MP3 format as ordinary music files. Though you could avoid this by staying away from audiobooks in MP3 format, there is a remedy to easily convert these into a compatible format with your iPod.

For starters, you should not upload an audible file that runs over five hours or is larger than 320 megabytes. By doing this, you will avoid problems with playback, such as pausing and stability.

If you happen to have files larger than the recommended amount, you may wish to separate them into smaller files. The majority of the audiobooks you download come in multiple files. You can combine these into one file so you can keep pace with the file you are listening to. This also makes the menu on your iPod much easier to navigate through. A good program to use for combining a batch of audible files in MP3 format is called MP3 Merger, which is available for free.

Converting MP3 Files for Audiobooks

Before the conversion of MP3 files takes place, the iTunes software must first be configured accordingly:

1.   Choose the "Edit" tab from the top left of the iTunes screen.

2.   In the next menu, select "Preferences."

3.   From there, click the "Advanced" tab.

4.   Now, select "Importing."

5.   Set your "Import Using" option to "AAC Encoder."

6.   Next, switch the setting to "Custom."

7.   Now, set the "Bit Rate" option to "64kbs." This is the typical setting for most audiobooks you will download.

8.   Set the "Channels" option to "Stereo." This repairs issues you may endure with mono files on your iPod.

9.   Enable the option "Optimize For Voice."

You can now add the files you wish to convert by importing them into the library like a typical audio track. Once they show up in the iTunes library, right click the file, scroll down the submenu, and choose "Convert Selection To AAC." This creates new versions of the files and stores them in your library.

When the files have been successfully converted, you can then remove the originals from your library, as they will be no longer needed. The new files can be found in the iTunes music folder of your computer. After opening the file, you will notice that they now have an .M4A extension. For them to be 100 percent compatible as audiobook files, they must be renamed to M4B.

You will then add the renamed files to the iTunes library. Right click one of the new files and choose "Get Info" from the submenu. This step is very similar to customizing your music tags, but the purpose here is to separate the audiobooks from your music. You may want to delete the fields for "Artist" and "Album," so these files are not recognized as music when playing on the iPod. You can use the tag editor and change the "Genre" of these files to something relative to the audiobook to make them easier to locate in iTunes.

From here, all you have to do is import the newly converted files to your iPod. Your manual conversions should play just like any pre-formatted audiobook you purchase and download. These new files will be recognized in the "AudioBooks" menu of your iPod, your playback spot will be saved like the Audible.com bookmark feature, and your audiobooks will not be mistakenly mixed and played into your music shuffle.

How to Rip Your Own Audiobooks

Audiobooks have been around for a long time. They were first introduced on cassette tapes and eventually made their way to CD. If you were already a fan, then there is a great chance that you already have a nice collection of audiobooks. The combined catalog from Audible.com and on the iTunes Store is impressive, but they may not have some of your old favorites. When this is the case you can easily rip your own audiobooks and sync them to your iPod.

To get the best experience out of your audiobooks, it is important to take note of these three elements: importing them in the proper format, sequentially combining all audio tracks, and making these files bookmark capable. Let us take a closer look at all three of these areas.

Importing Your Audiobooks

Your iTunes application is set by default to import music in an AAC file format. Since audiobooks and other recordings of the spoken word do not require the same level of sound quality as music, you can save a bit of space on the hard drive by changing the original format. To do this, open the "Preference" menu on iTunes and then click the "Importing" tab.

If you want your files in the standard AAC format, select the "AAC" encoding option. If you want to optimize your hard drive, you can choose "MP3 Encoding" and apply the steps mentioned above to avoid the problems with bookmarks.

Next, you will select the "Settings" menu and choose the "Custom" tab. Select "64kbps" from the "Stereo Bit Rate" menu, as this is the standard rate for books you download from Audible.com and the iTunes Store.

NOTE: After ripping all of your audiobooks, you will want to reconfigure these options to their original settings. Ripping your music at this bit rate will falter its quality.

Combining Audiobook Tracks

Most audiobooks are composed of many different files. It is much better to combine all of these into a single track to make the audiobook more accessible on your iPod. To combine audio tracks while importing your audiobook, you first need to insert the disc into your computer. The iTunes software will generally detect the contents of an original audiobook just like a music disc. Select "Edit" in the top left of the iTunes screen, then choose "Select All" from the following window. This will highlight all the files of your audiobook.

Next, select the "Advanced" tab. In the following menu, click "Join CD Tracks." The iTunes software will then display the files of your audiobook with a bracket that indicates they have combined.

Rip Your Audiobooks for Bookmarks

The final step in effectively ripping your own audiobooks is to make them bookmark capable. While the concept is similar to the file conversion we just covered, the steps you will take are different. Mac users have the aid of many Apple scripts designed by independent developers that make bookmarks easy. A Windows user can get the same results by importing their audiobook CD and right clicking the files. From there, choose "Show Song File," then modify the file extension from M4A to M4B.

Your audiobooks are now set to be synced to your iPod. Arrange them in playlists like you would do your music, or leave them just as they are.

Making Your Audiobook Cassettes Digital

By now, you know that any audio file imported to the iTunes application can be synced to any iPod. In the case of audiobooks, you can also import these using various forms of compression whether it is AAC or MP3 formats or with the use of an encoder.

Audiobooks on cassette tapes can be ripped, digitalized, imported into the iTunes application, and then synced to your iPod. As this program specializes in handling uncompressed audio files, even tracks from your audiobook may be import with little to no decline in quality.

To digitalize and rip music from cassette tapes, you need an audio card that contains a two-channel level input. The line level outputs of the cassette player must be connected to level inputs of the audio card. The tracks can then be recorded as a single file or individually by way of an audio editing program. I recommend Audacity for this task as well.

In many cases, there will be a sound card already installed on your computer. Sometimes, the exterior speakers you use will have line level inputs as well. While they can aid in the process of ripping your audio from cassettes, there might be an issue here.It is strongly suggested that you install an additional component, such as Pro-Audio Card, which will give you much better results in the end. This type of sound card can be found online and at many record stores.

Your iPod On-The-Go

The beginning stages of life with an iPod will require lots time on the iTunes program. This is where you will import music, create playlists, and sync to the iPod. From there, you have enough music to get you through the day and then some. iPods are commonly used for recreation or in the car, and have become an everyday workout partner for many.

Since your iPod is soon to become your little traveling companion, it is imperative to get familiar with the special little feature — on-the-go playlists. As convenient as this function may be, several of my fellow iPodders have no clue of how to work it, or that it even exists.

Though often underated, this feature is great for the iPod fanatic constantly on the go, hence the name. It allows you to create new playlists directly for your iPod, saving you the trouble of accessing the iTunes program. Once the playlists have been created you have the option to rename and save them in iTunes.

The on-the-go feature works with music and podcasts as well. Podcasts will be listed in reverse order on the iPod. If you want to listen to them in sequence, add them to your on-the-go playlists.

Create On-The-Go-Playlists

Step 1: From the main menu of the iPod screen, select "Menu."

Step 2: To add content from your favorite artists, select "Artist." To add complete albums select "Album." To build by individual tracks, choose "Songs."

Step 3: Use the click wheel to scroll through your options. After finding the artist, album, or song you want to add, press down "Select." Your selection will then blink, indicating that it has been added to the on-the-go playlist.

After building the new playlist, navigate back to the main menu by pressing the "Menu" button. To find the playlist you just created, click "Music" from the main menu, access the "Playlists" menu, then scroll down to "On-The-Go." Next, click "Play/Pause" and the songs will begin.

NOTE: While you can go back and add to your on-to-go playlists at any time, you can only hold them one at a time on your iPod. The best thing here would be to save them in iTunes so you can use them later. To delete an on-the-go playlist, navigate down the menu and select "Clear."

Smarter Shuffling System

The shuffling feature ironically came about with the introduction of the iPod Shuffle. While disk space on the original model was nothing to gloat about, shuffling has since graduated to more powerful versions where thousands of your songs can be mixed and played at random.

The shuffling feature on your iPod is one you may use rather frequently. This is a hassle free way of playing your music when entertaining friends, cleaning the house, or when you get tired hearing the same playlists. As popular as the iPod's shuffling feature is, many users often wonder whether it is really random.

These days, you may find several complaints in online forums and blogs concerning the iPod's shuffling feature. Some users are reporting tales of songs being repeated and other issues such as not hearing their favorites or hearing consecutive songs by the same artists. This has led a number of people to believe that the randomness of the iPod is not what it is made out to be.

When breaking this issue down in definitive terms, you will find that random plays does not necessarily mean songs will not be repeated. A random shuffle is simply that — songs are generated in no particular planned order.

So what does this group of unsatisfied iPod shufflers really want? They desire a shuffle list that is thoroughly mixed up and plays a vast amount of fresh music before a single song is repeated. Since the shuffling system is random, there is no way to get exactly what you want. However, there is a way to get better results out of your iPod's shuffle mode. The answer lies in smart playlists. This is the best way to command both the iTunes program and your iPod to do what you want.

We covered smartlists and how to create them early in the book. Let us now work on some advanced methods for shuffling. Start by clicking "File" at the top left of your iTunes display screen. Navigate down the menu and select "New Smart Playlists." When the menu opens, you will notice "Limit to" with other options and boxes surrounding it. Here is where you can set your playlists for a certain amount of songs. By making use of the limit control, you set the standards for what type of songs will be played in the shuffle as opposed to iTunes selecting them.

An alternative to this is getting the aid of third-party software. One such program is called iTunes Randomizer. This application is free to use and specifically written for the Macintosh operating system. It is very unique in that it lets you randomize by values such as minutes, hours, and days.

Regardless of the strategies you employ, having 100 percent authority over your shuffling feature may never be possible. For now, you can prompt better results by exercising a bit of creativity and patience.

Case Study: Tee Morris

Tee Morris

Author, Podcaster, Actor

Actually, I stumbled on writing by accident. It's been something I have always done for fun, but never took seriously. It was when I was on a less-than-stimulating temp job and I was passing time in between assignments by playing in an online chat room.

I was playing the character Rafe Rafton, my character from the Maryland Renaissance Festival; and that is when I met Lisa Lee, under the character alias of Askana Moldarin. We started up an e-mail role-play; and about three months into it, I was spending less time in the chat room and more on the word processor. We realized at the halfway point of *MOREVI* "Hey, we are writing a book!" I did the homework on publishing options, researched Dragon Moon Press, and in 2002 *MOREVI* was published. Since then, I have had a lot of terrific accomplishments as a writer; but I still have a lot of stories to tell and a good way to go before I am a Terry Brooks or Tracy Hickman.

It was my dream since high school to be an actor and that is where and when the actor's bug bit me and bit me hard. I studied theatre in college and sought out roles that people might not readily see me in. The one thing I share in common with Harry Potter's Daniel Radcliffe is the role of Alan Strang in Equus. Intense role, and for Radcliffe another fantastic opportunity.

I also enjoyed a few memorable moments in the spotlight, my biggest one being a small walk-on role in an episode of "Homicide: Life on the Street." The Maryland Renaissance Festival was the most fun (outside of a 1998 production of "The Comedy of Errors" with Vpstart Crow Productions) I have had as an actor, and that opened an unexpected door for me — the role of published author.

As a writer, podcasting has been a real advantage in getting my name and work out into the public eye. I have done a lot of various promotions since 2002 to let people know about *MOREVI*, but podcasting has introduced me to an international audience and continues to do so. When Billibub Baddings entered the podosphere, *MOREVI* returned to the spotlight again as people wanted to hear more from Tee Morris. I decided to go back to its original podcast and remaster it from the original AIFFs I used in 2005. "MOREVI: Remastered" has been a real challenge, and a heck of a lot of fun! Then there is "The Survival Guide to Writing Fantasy" which is a marketing and self-promotion podcast for writers who get little to no consultation on the matter. All these projects led to *Podcasting for Dummies*, and I still get a great amount of opportunities through my affiliation with Wiley Publishing.

Case Study: Tee Morris

Podcasts have also been a heck of a lot of fun to produce. Sometimes, I have more fun producing segments for others than full shows for myself. The community is very supportive in podcasting, and I feel very fortunate to be a part of such a fantastic group of trendsetters.

Simply put, enough people bought "Podcasting for Dummies" that Wiley said "Let's see if we can do this again!" I was all set to go for round two, Evo was not so ready, and Ryan was there to say "I'll help out." That is the short version...

Now the long...

By the time *Podcasting for Dummies* came out, a lot of changes had occurred, especially with enhanced and video podcasting. Even with "Podcasting for Dummies: The Companion Podcast" giving the book an extra bit of promotion and audio addendum, "Podcasting for Dummies" was not considered a potential hit for Wiley. That was before it went into a second print within its first month.

Both the book and the podcast were runaway successes for the For Dummies people, so much that feedback started trickling in asking about an update for the title. It was needed between 2005 and 2007, so much had changed. Initially we (Evo and I) thought a second edition would be fine, but Wiley wanted a second book... and a second edition. Currently, I am working with Chuck Tomasi on the second edition (still geared for beginners), and *Expert Podcasting Practices for Dummies* (which focuses heavily on production, professional grade recording software and equipment, and applying podcasting in a business setting) is now in bookstores everywhere...and hopefully, selling out.

I have always loved the fantasy genre (Terry Brooks is a hero of mine). *MOREVI* did not really start out as a fantasy so much as it evolved into one, and I still love returning to that universe Lisa and I created. When I sum up *MOREVI* for people curious about it, the one-liner makes people smile: It's *Crouching Tiger, Hidden Dragon* meets *Pirates of the Caribbean*. As much as I love the Chinese culture, fables, and folklore, I also love my science fiction and fantasy to have a touch of the swashbuckler in them. I am particularly fond of the movies of Errol Flynn and The Three Musketeers with Michael York. I grin like a kid when I watch Star Trek: The Next Generation's "Qpid" where Picard and the crew are turned into Robin Hood and his Merry Men. Brilliant! When I wrote *MOREVI*, it was the best of both worlds and a lot of fun to write. *Legacy of Morevi* took a slightly darker turn, but I'm thinking the tone will be lightening up in *Exodus from Morevi*. I do not want Rafe and Askana to get too serious, you know?

Case Study: Tee Morris

Speaking of not being too serious, I wanted my next book to take an entirely different turn, to show others (and myself) that I was not just a one-trick pony. It was a song from Leonard Nimoy (no kidding) that inspired *Billibub Baddings and The Case of The Singing Sword*, a book that was another "best of both worlds" for me. I'm also a fan of mobster movies and the true stories of gangsters like Capone and Gotti. So *The Case of The Singing Sword* was my chance to pay homage to fantasy and detective novels, and then give them both a playful ribbing. The book was a bit like *Podcasting for Dummies* with my publisher. Dragon Moon Press was not quite sure what to make of a dwarf detective in 1929 Chicago, and did not really expect it to go anywhere. It wound up becoming one of Dragon Moon's best selling titles, and is still selling nicely with the success of the podcast. Next for Billi is a baseball-based mystery entitled *The Case of The Pitcher's Pendant.* I think this one is going to be a lot of fun for readers.

When I had completed *MOREVI* in the summer of 2005, I had this gear sitting in my office and was trying to figure out what I was going to do with it next. A dangerous question, now that I reflect on it, because this was before I had launched this show, "Podcasting for Dummies: The Companion Podcast," and others. Writers, even in the fall of 2005 were stepping into the podosphere with writing podcasts, and as I had just written *Podcasting for Dummies*, I was trying to figure out what to bring to the plate that other writers were not. Character development? Worldbuilding? Plot structure? All these topic were either covered or were being covered, and while I could go on and bring in my perspectives I wanted to produce something unique.

Then I got an e-mail from someone who had attended Westercon (the SF convention where I had premiered *Legacy of Morevi*), thanking me for talking to them about self-promotion and marketing. The e-mail made me recall an article I read in the local paper about the book market: In one year, an average of 170,000 books is published. That is across all genres, from academic to corporate press, and from authors new and established. I started to think about how in those 170,000 published in 2005 I wanted people to buy two: *Legacy of Morevi* and *Podcasting for Dummies.* I thought to myself, "Well, I know a few things on how to market and promote my work, but what about the 169,998 other books?" From this, "The Survival Guide to Writing Fantasy" started to take shape.

"The Survival Guide to Writing Fantasy" is a podcast for writers (of any genre) who want to know more about how to market their work and themselves. In the two years I have done "The Survival Guide" I have broadened the scope of the show to cover topic that

Case Study: Tee Morris

maybe an author does not always consider or may prevent the writer to do what they are supposed to do: write. Past shows have included "Dealing with Distractions," "Taking Care of Yourself," and "Tough Love: Coping with Bad Reviews." This past year, I have been inviting more people on "The Survival Guide," giving interviews with people in the industry who are either sharing their approach to marketing or finding out what they do to stay writing. I am now in 2008, the show is on a monthly schedule (on account of other projects, including "Morevi: Remastered"), and the fan base for it is solid. I have been very happy with how the show has matured and held up over the years.

I still remember my first iPod commercial (2001) of a guy syncing his iPod up and grooving to the playlist he had just created. I also remember looking up the cost of one and thinking, "You've got to be kidding!" I believed the iPod was gong to be Apple's next G4 Cube or Newton until I listened to a friend's iPod on the way to a concert. It was kind of nice to be able to carry around that much music, and my car was full of CDs and cassettes, so in 2003 I went on eBay and picked up for myself a used 2nd generation 10GB iPod (complete with someone else's name engraved on it). That is when I understood.

It is not so much its built-in features but what is out there that can help you expand on its possibilities. When I was searching out accessories, I found a vendor online called H2O Audio (**http://h2oaudio.com**) and they make outdoor and underwater cases and gear for iPods. I picked up an armband and watertight case for my iPod (guaranteed up to 10 ft.), and I now have music for when I swim. Coupling that with the stopwatch feature included, I can now keep time with my swimming and see how many laps I have completed and if I am improving or not.

Then there are the pictures and videos of the 5th generation iPods. It is really, really nice to be able to have those on call when I want to chill out in an airport or just pull up a few fun images of my kid. The iPod has a lot of other functions going for it that I have not taken advantage of (yet), but it is a valuable part of my daily routine and I feel naked without it.

When it comes to other brands of MP3 players, I am familiar with several. There is the iRiver which, at one time, I had high praise for...until their models (starting with the T-series) started catering to the Windows crowd. The players refused to interface with the Mac so I looked at other MP3 players. Another one of note, while more of a recording device than an MP3 player, is the Zoom H2. It works as a multi-purpose tool (SD card reader, recorder, player); and while not as compact as an iRiver, it is portable. Still, my heart belongs to Apple's iPod.

Case Study: Tee Morris

If you have never owned an iPod and want to try one out, I would recommend the Shuffle for two reasons: its low-cost investment and its ingenious design. It has all the sound quality of an iPod so you can get an idea of how it is going to sound, but on account of its design, you not only get the sound of the iPod but you get that Apple "We try to think of everything" touch by being able to clip that puppy virtually anywhere. It is light, convenient, and easy to figure out. If you really dig it, skip the Nano and go for the Classic iPod. More bang for the buck. Until the iTouch doubles (if not, triples) their capacity, I think I will be sticking with the Classic models.

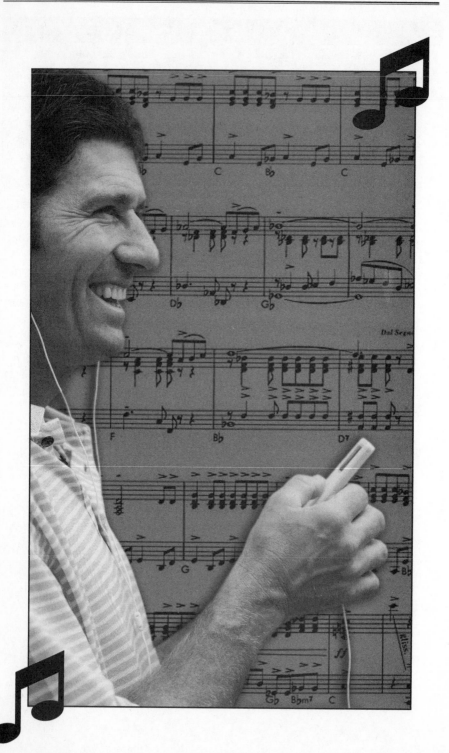

Advanced Work with the iPod — Part 1

iPod Playback and Sound Options

Most people will tell you that an iPod comes ready to play with exceptional sound quality. The bass capability and crisp voice reproductions on this device will put your once cherished Walkman to shame. The three levels of the make-shift equalizer you are accustomed to will seem rather primitive compared to the sound options you have with an iPod.

One quality playback option you have for your music is the crossfade. This is great for playing the latest tunes and gives the feel of a professional radio show with you playing the role of disc jockey. Here is how to set up crossfade playback within the iTunes program:

1. Click "Edit" from in the top left of the iTunes screen.

2. Next, click "Preferences."

3. In the Preferences menu, click "Playback."

In this menu, you will notice these three options: "Crossfade playback," "Sound Enhancer," and "Sound Check." Mark the checkbox beside "Crossfade playback." You can then make adjustments to the level in regard to time from one to 12 seconds.

More on Sound Options

You can think of the Sound Enhance function as the surround sound system control for your iPod. By marking the box beside this option you can adjust the level from high to low. The sound enhancer can provide a more crisp and clear sound. This option gives the best result when listening to your music on a set of headphones.

I would not recommend the crossfade playback for someone who likes to listen to classical artists such as Beethoven and Mozart. A quality option for music of that nature would be "Sound Check," as it automatically adjusts playback of a track to the same level for a more natural effect. The first time you enable this option, iTunes will automatically analyze the tracks to determine what level they should be played at. The trouble with this is the different levels and instruments used to create a song. Some are high and others are low, which means the sound check options could work well for one track and produce poor quality on the next.

NOTE: Playback settings must be adjusted and applied to playlists on an individual basis.

The Equalizer in iTunes

Many of us are so into our music that we seldom notice the flaws within. The secret is that no matter how good the tunes sound on your radio, they can never truly compare to the originals.

Travel back to the old days for a second. Remember recording that tape from a friend who dubbed it from another friend. It was rather apparent that the quality of the music declined with each recording.

Equalizers are used to combat issues of playback. You can find them within your home stereo system, on your car radio, and even on your iTunes software and iPod itself. Unknown to most, there is actually an equalizer built into the iTunes that allows you to minimize recording flaws and gives you playback results that mimic the original production.

Playing with the iTunes Equalizer

The mere thought of tampering with the equalizer may appear somewhat intimidating at first glance. In actuality, it is pretty straight forward once you are familiar with all of the functions and how they work. To access the equalizer, click "View" at the top left of your iTunes screen. The first control you notice will be a "Preamp" slider. This will act as the master volume control for your music. For starters, I suggest leaving this adjustment alone and using the volume control on iTunes or your portable medial player.

The iTunes equalizer gives you 10 different frequencies to play with. Each has its own slider that allows you to make adjustments to a certain range of frequency. While some programs may have more bands on their equalizer, they all perform a similar function, which is to apply different types of sound to your audio. The 10 bands found of the iTunes equalizer are designed to control three different types of sound: Bass, Treble, and Mid-Range. The sections below detail each area.

Bass Control on iTunes Equalizer

For those of you who not only like to hear you music, but desire to feel it thumping in your ears, the iTunes equalizer has you covered. Adjustments of your bass will be from the first two sliders on the equalizer. For best results with bass, I suggest using high-amped speakers, such as subwoofers. These adjustments will also be reflected through your headphones or ear-buds.

You may not notice these adjustments on basic speakers that came with your computer (they are not designed to playback extreme bass).

Treble Control on iTunes Equalizer

The final two sliders found on the equalizer are responsible for treble. The treble effect brings out the high pitches of instruments and even voices. Treble controls mostly the harmonic tones of music. After increasing the treble, your music may seem louder, and the words clearer as it works against the bass. Decrease the treble and music sounds hollow.

Mid-range Control on iTunes Equalizer

You will find the mid-range frequencies right in the middle; they are controlled by six sliders. Most of the audio in any music is found in the mid-range area, particularly voices and instruments such as drums, pianos, and guitars. While you will use these frequencies to make the music more crisp and clear, these sliders can be adjusted to give you some quality bass on the low end as well.

Making Adjustments

In most versions of iTunes, the equalizer will be enabled by default. If not, all you have to do is mark "On," which is located just above the Preamp slider. From there, you can select your presets. Here is a list of all the available presets on iTunes:

♫ Acoustic   ♫ Bass Booster

♫ Bass Reducer   ♫ Classical

♫ Dance   ♫ Deep

♫ Electronic   ♫ Flat

♫ Hip-Hop   ♫ Jazz

♫ Latin   ♫ Loudness

♫ Lounge   ♫ Piano

♫ Pop   ♫ R&B

♫ Rock   ♫ Small Speakers

♫ Spoken Word   ♫ Treble Booster

♫ Treble Reducer   ♫ Voice Booster

Your presets will adjust different frequencies to make up for lost quality that might occur in your headphones or speakers. In a sense, they give you direction on how to adjust the equalizer according to your music. You can also choose manual, which allows you to make adjustments to the sliders on your own.

The bass booster and treble reducer clearly increase and reduce those two frequencies. The other presets you find are not actually true to their label. Many of them are labeled by genre and are designed to adjust music accordingly. The main function of these presets really is to increase or decrease volume, particularly in the areas of bass and treble.

The best way to determine the appropriate preset for music is giving the equalizer a try. You may first want to listen to the music with the equalizer both on and off to determine whether or not you even need to tamper with it. Test the music and see how your selections sound on iTunes, through your headphones, or on a set of external speakers. If the speakers being used are not the greatest, you may want to enable the bass reducer to longevity.

How to Make Your Own Presets

Setting your iTunes equalizer is the first step in making your own presets. From there, you can adjust the sliders to your liking, as mentioned above. When your bass, treble, and mid-ranges are set, click the dialog box of the sliders and choose "Make Preset." This will prompt an empty box in which you can give it a name. Then, click "OK" and your preset will be added into the default listing and can be applied to any of your songs or playlists. You can rename or delete your custom presets at any time. If you decide to delete a custom preset later, iTunes will ask if you want to remove the setting from all tracks.

Presets for Individual Tracks

The easiest thing to do in regard to the iTunes equalizer is to customize your presets for each and every song. The problem is song quality. The preset for your favorite hip-hop may not match up well with the heavy guitar and drums of a rock track. Fiddling with the sliders for your bass and treble is an option, but this could end up having an adverse effect on the quality of voice and certain instruments.

In this instance, you can designate presets for individual songs that cater to the different sounds and instruments they have. To begin, you will highlight a track and right click it or hit "File" at the top left of the iTunes screen. In both of these menus, you will find "Get Info." From there, click the "Options" tab and choose a preset from the menu.

If you want to choose a single preset for a number of songs, click "Edit" at the top left of the iTunes screen and choose "Select All" in the following menu. This will highlight the designated songs. Next, right click the items and follow the steps above to choose your preset.

The iPod Equalizer

When away from the computer with no access to iTunes, you can use the equalizer on your iPod to adjust the sound of your music. This version is similar to the one found on iTunes, but is not as extensive. For instance, you will find the basic presets, but will not be able to adjust the individual bands.

To access the equalizer in your iPod, choose "Settings" from the main menu. Select "EQ" on the next screen. Scroll down the equalizer screen to find the setting of your liking, then press "Select."

Keep in mind that the chosen preset will be assigned to all the songs on your iPod. To change these settings, select "Off" on the EQ screen.

The EQ setting on your iPod is a great luxury when you are on the go with the device. However, there are three things you will not be able to do with the iPod equalizer:

1. The iPod will not use equalizer presets you made in the iTunes program.

2. Custom presets cannot be made on an iPod, such as adjusting the bands and naming the setting.

3. If a song is assigned a specific preset in iTunes, it will keep the same form on the iPod, even if you change the settings.

NOTE: Songs with presets from iTunes can be played temporarily with EQ settings from the iPod. Here is how to do it: As the designated song is playing, tap the Menu button until you have been taken back to the Main Menu screen.

Select settings.

Choose EQ from this menu.

Select the EQ setting you wish to apply to that song.

Select the EQ setting you wish to apply to that song.

The song will then reflect the new adjustments you set in place. When you play the song again, it will revert to the presets assigned in the iTunes program.

iCal

The calender function, most commonly known as iCal, is a digital calender created by Apple for the exchanging of data. Though the calender function is apparent, iCal has many more capabilities that allow you to stay organized and on schedule. With iCal, users can perform business tasks, such as sending meeting requests via e-mail.

Here is a rundown of some of the features you can enjoy with iCal:

♪ Stay up-to-date with appointments, special events, and your personal calender. View your schedule and upcoming activities on a daily, weekly, or monthly basis.

♪ Manage and view multiple calenders at once from a single display to identify scheduling issues in a timely fashion.

♪ Share calenders online with your Mac account.

♪   Subscribe to different calenders and stay up-to-date with your work schedule and other important events.

♪   Transmit typical e-mail invites to e-mail addresses on any Web server.

♪   Stay current with the built-in To Do List management function.

♪   Alerts of upcoming events via e-mail, audio, or screen.

♪   Use the swift search engine to find any name, task, or event you may be looking for.

This section will thoroughly explain the benefits of configuring your iPod as a digital secretary by using the iCal application. From there, we will cover the steps it takes to synchronize everything with the Apple application, iSync.

iCal Requirements

Macintosh Operating System X version 10.2.3 or later is required for using iCal.

NOTE: Syncing iCal data to portable devices, such as iPods, cell phones, PDAs, and other Macintosh products requires iCal version 1.5.5. A Mac membership or WebDAV server is required to publish calenders online. iCal and all of its components can be downloaded for free via the Apple Web site.

The iCal Interface

Let us begin by discussing the iCal interface — what you will see when first starting up. The main focus of the screen is your calender view. You can choose for this view to be on a daily, weekly, or monthly basis by accessing the three buttons at the bottom of the display. I recommend the monthly calendar view, as it allows you to quickly reference days of the week, much like the one you may have at home. This is also an easy way of browsing to add items of importance in the future. iCal allows you to display up to three months at a time. You can even adjust the size with the easy to use horizontal bar, located just above the calendar.

In the top left corner, you will find a "Calendar Listing." The default versions are for "Work" and "Home." To add a new type of calendar, simply click the "Plus" icon, located in the lower left corner of the screen.

Mac subscribers have the advantage of publishing their calendars on the internet to share with others. This is pretty easy — simply highlight the designated calendar and click " Calendar" and then select "Publish."

Just below the primary iCal window, you will find a box for the search feature. Use this function to locate any appointments or dates you might have coming up. This is easier than flipping from month to month.

Next, we will cover one of iCal's most complex, yet useful features. The "Details Drawer" could be considered iCal's control panel, as this is where you can set all options and configurations. Here, you can set details for entire calendars as well as appointments. You can customize your appointments by detailing whether it is an all-day event by setting the length of time. This is very convenient for a work or school schedule to remind you when a break is over between classes.

With iCal, you can also assign a certain status to particular events. This is helpful toward your business endeavors, as you can determine when an appointment has been confirmed, pending, or cancelled. You will never forget an important appointment or birthday again with iCal's alarm notification features. These settings can easily be synced to your iPod on those occasions when you are away from the computer.

One of iCal's most beneficial features is the "To Do List." We could all use one of these in our lives. Being able to keep this list in your pocket makes it even better. The "To Do List" allows you to set special tasks for particular calenders with designated dates, and even gives you Web site URLs to go along with it.

In addition to creating and publishing your own calenders, you can receive calendars from friends and family by subscribing to their creations from the Apple Web site or a wide range of third-party Web sites offering calendars. Below are two helpful Web sites to pull iCal calendars from:

1. **iCalShare.com** — This site offers a variety of calendars with many different genres to choose from. Select from sports schedules, TV Guide menus, and more.

2. **ApplesiCalListing.com** — All calendars found on this site are provided to you by Apple.

The selection is not as frequently updated as those you will find on iCal, though you are sure find a few calendars that are worth using.

iSync

After playing with the iCal application for a while, you will probably want to transfer the data to your iPod. This is where iSync comes into play. Together, iCal and iSync make managing your important data simple, allowing control directly from your iPod. Similar to iCal, iSync can be downloaded via the Apple Web site.

The iSync interface contains a designated button for each portable device it is compatible with: Mac, iPod, cell phone, and PDA. You can configure each device individually by clicking on its icon.

How to Sync iCal with iSync

1. First, connect your iPod to the computer with the USB cable.

2. Next, open the iSync application.

3. When the program opens, click "Sync Devices." iSync will then automatically scan for devices. The "iPod" icon should then appear in the "Add Device" window.

4. Double click the icon and add it to the iSync menu. This menu will display options that allow you to select what data to sync to your device.

5. If you want iSync to automatically sync the information when the device is connected, mark the "Automatically synchronize when device is connected" box beside it.

6. Next, select the iCal calendar you want added to the iPod.

7. Click "Sync Devices."

How to Sync iCal with iTunes

Now that you are a bit more familiar with the functions of iCal and iSync, let us work on setting everything up:

1. First, connect your iPod to the computer via the USB cable.

2. Next, open the iTunes application.

3.   Select the iPod icon on the left side of the iTunes screen.

4.   Click on the "Contacts" tab.

5.   To enable calendar syncing, scroll down and check the box beside "Sync iCal Calenders."

6.   Next, click "Apply" and the iCal sync will begin.

NOTE: After you begin using the iTunes program to sync iCal calenders, you will no longer be able to use iSync to sync that content. iTunes will assume all responsibilities and cancel out the iSync application. You should also know that iCal calendars that have a colon (:) in the file name cannot be synced to your iPod.

How to View Your Calendars

After the iCal calendars have been synced to your iPod, you should view them and make sure everything went accordingly. To do so, you will need to follow these steps:

Step 1: Click the "Eject" button located next the "iPod" icon in the iTunes display.

Step 2: Now, on the iPod, select "Extras" from the "Main Menu."

Step 3: In the next screen, you will choose "Calendars." Select "All" to view all calendars or simply highlight one by name and press "Select." The iPod will display calendars marked by flags, indicating events or scheduled appointments. To view a certain event, rotate the click wheel on your iPod, navigate to the date, and press "Select." This will show you a list of all events pertaining to that date. You can then choose the event to view of its details.

Contacts in Your iPod

The iPod eliminates the hassle of taking that big, clunky event planner every where you go. Many people have an address book or piece of paper taped to the wall with a list of emergency phone numbers and contacts. If you have this information stored on your computer, transfer it to your iPod to make things a lot easier. The ability to sync contact information to an iPod is a great perk for both Mac and Windows users.

Syncing Contacts from a Mac

The process of syncing contacts to your iPod is very similar to adding iCal calendars. Let us review the steps:

> **Step 1:** First, connect your iPod to the computer via the USB cable.

> **Step 2:** Next, open the iTunes application.

> **Step 3:** Select the iPod icon on the left side of the iTunes screen.

> **Step 4:** Click on the "Contacts" tab.

> **Step 5:** To add your contacts, select "Sync Address Book contacts."

> **Step 6:** Next, click "Apply" and the syncing of your contacts will begin.

NOTE: Similar to your iCal calendars, iTunes will handle all syncing duties of adding content, overruling the iSync application.

Syncing Contacts from Windows

Contacts from your Microsoft Outlook and Outlook Express

applications can also be synced to an iPod with iTunes. Just follow these directions:

Step 1: Open the iTunes program.

Step 2: Next, connect your iPod with the USB cable.

Step 3: Click the "iPod" icon when it displays to the left of the iTunes screen.

Step 4: In the iPod menu, click the "Contacts" tab.

Step 5: Enable "Synchronize contacts from" by marking the checkbox beside the text.

Step 6: From this menu, you will choose a program to sync from: "Outlook," "Outlook Express," or "Microsoft Address Book."

Step 7: Click "Apply."

This is the end of the syncing process for Outlook Express users. Outlook users need to take note of the following section to instruct the iTunes program to access the data.

After clicking the "Apply" tab, a dialog box will appear advising you that a program is attempting to gather data from Outlook. For this to happen, you must mark the "Allow access for" checkbox. From here you will click "Yes" to close the dialog box. Lastly, click "OK" to close the iTunes preferences and initiate the final part of the syncing process.

NOTE: You must allow access to the Outlook application every time you sync contacts from there to your iPod.

iPod Games

When you have played out your music and grow tired of watching the same movies, the recreational magic can be rekindled with games on your iPod. Apple has a variety of games available on the iTunes Store, from classics like Pac-man to new favorites such as Texas Hold 'Em and LOST.

Buying and Syncing Games from the iTunes Store

After following these simple instructions, you will be playing games on your iPod in no time. Let us begin by detailing the steps of purchasing games and then move on to the syncing process.

1. First, open the iTunes program.

2. Next, click on the icon for the iTunes Store.

3. Locate and click the "Games" tab on the left hand side of the iTunes Store. A menu of all the featured games will display.

4. Browse the menu to find a game of your liking. Several of them have demo versions that allow you to test games before purchasing them. After finding the perfect game, click on "Buy Game" and purchase with your credit card or PayPal account.

5. Connect your iPod to the computer via the USB cable.

6. Click the "iPod" icon when it shows up on the left side of the iTunes screen.

7. In the iPod menu, select "Sync Games."

8. Next, you have the option of syncing "All Games" or "Selected Games." Mark the appropriate box.

9. Click "Apply."

NOTE: The games available on the iTunes Store are only compatible with the iPod Nano, iPod Classic, and iPod Video. Furthermore, the iPod video has its own line of games that may or may not be compatible with the iPod Nano and the iPod Classic. Visit the iTunes Store to get a full list of games and to determine what model they are available for.

Cheats for iPod Games

Not only can you play a slew of games from your iPod, you also have access to the cheat codes that enable you to master them faster. The Apple company released exposed cheats for two of their most popular iPod games: Texas Hold 'Em and Vortex. These codes allow you to start out with much more money, enable you to access hidden stages, and much more.

Texas Hold 'Em Cheats

To access cheats for this game, you will go to the "Options" menu on your iPod screen and select "New Player." You can enter a code using the letter selection tool on the iPod. When the cheat has been entered, your iPod will display a checkmark. From there, hold down the "Select" button and wait for the code to activate. You will know the code has been activated when the text, "Secret" appears on the screen.

LIST OF TEXAS HOLD 'EM CHEATS

ALLCHARS: Allows you to view secret players

BARTUNES: Allows you play the iTunes Bar Tournament

THREEAMI: Allows you play the Apple Conference Room Tournament

BIGROCKS: Allows you to play the Stonehenge Tournament

PLAYDOGS: Allows you to play the Dog Tournament

SPACEACE: Allows you to play the Futuristic Tournament

Vortex Cheats

To access cheats for this game, you must bring up the "Main Menu" before starting it. You will then notice the option "Personal Info." From this menu, select "Rename" and enter a code using the letter selection tool. Then hold down the "Select" button and wait for the code to activate.

LIST OF VORTEX CHEATS

ME_PAZ: Activates laser power-up capability

I_GUNZ: Activates the gun power-up, enabling you to blast bricks one by one

FORSIX: Activates additional playing time, giving you 24 lives

_PWR_B: Activates the power ball, enabling you to blast multiple bricks at a time

Other iPod Games

Though you will find an extensive list of games in the iTunes Store, there are others available from third-party Web sites. Most of these suppliers will provide you with software that allows you to convert the games and make them compatible with your iPod. Since these games are not licensed by the Apple company, you should certainly take extreme precautions when choosing to sync them to your iPod, as there is a risk of contracting spyware and viruses from these sources.

Case Study: Christy Hurst

Christy Hurst

Quality Assurance Analyst

My job — it is basically a software testing position for Central Transport International's intranet and customer Web sites. When I graduated high school I wanted to choose a career path that would be both interesting and challenging.

I have always had a desire to work with computers. I loved video games, the Internet, and working with different applications. The way that computers are able to make our lives easier, create fun and exciting things to do, and the vast amount of knowledge you can find from "surfing the net" had peaked my interest and I knew it was something I wanted to learn more about. Technology itself has always fascinated me and the thought of having an opportunity to go to school to learn about the inner workings of a computer, how it is put together, its "guts" and all its glory got me thinking, "Hey, I could do this the rest of my life!" That thought alone got me hooked and so I signed up for my first class in Computer Science. Walking away with that degree enabled me to become skilled in computer hardware and software and to this day, I am still astonished by how much I still love learning about technology and all it can offer to us.

Working out is my one true passion. I love to lift weights, run, roller blade, hike, bike ride, and play sports. Basically, anything physically challenging is my favorite thing to do. I love playing interactive video games, surfing the net, and my computer in general. Going out to eat is a favorite of mine, too. Hanging out with friends and either going out to a bar, or just chilling in front of the tube watching movies will always be one of my favorite things to do. Driving around town is also a lot of fun for me because I can crank up the tunes. I love to listen to music and dance. My music keeps me sane. Learning (i.e. reading as much as possible) is still a favorite pastime for me also. I would go to school for the rest of my life if I could! That is me in a nutshell.

My initial reaction to the iPod was, "For real? It's that small?!" My first model was the iPod Mini. It sounds cheesy but I love the fact that I can put my CDs on my iPod since I

Case Study: Christy Hurst

have hundreds of them. The best "feature" is probably the ease of being able to scroll through the songs I have and create my own playlists for easy listening.

I am able to get the most enjoyment from my iPod at the gym and anytime I go out for a walk, run, or bike ride. When I lay out in the sun is also another time where my iPod comes in handy. It is like my little friend I cannot live without!

I have between 350 to 400 songs on my iPod. I never listen to just one artist. I put it on shuffle so that I can get a mixture of all the fast paced punk rock music that keeps me moving.

I would truly love to be able to listen to FM radio on my iPod. Sometimes, I just do not feel like listening to the songs that I have loaded, so a radio option would be heaven on earth!

If I had to recommend a model for someone looking to buy their first iPod it would be the iPod Nano — it is so cute!

9

Advanced Work with the iPod — Part 2

Locking Your iPod

The iPod has advanced in such a way that some models give the feel of a mini Mac or PC. With the proper tools, you can secure your PC from intruders. You can also configure your iPod for security to keep your content confidential.

The screen lock function enables you to assign a four-digit code to your iPod. This completely denies access to the device, and can only be unlocked by entering the code. The screen lock feature is exclusive to the iPod Nano and the iPod Video.

Do not confuse the screen lock with the hold switch. The hold switch will only prevent the iPod from being powered by accident. It is not made for keeping the contents of your iPod secure.

Truth About the Screen Lock

The screen lock is very useful, yet limited in a sense. It has the power to block intruders from the interface of your iPod, but it is not capable of encrypting your content. If your iPod is connected to a computer, it can be managed while in disk mode, even if the screen lock has been set in place.

Setting Screen Lock Combinations

To use this feature effectively, you must first set the combination. Just follow these steps:

Step 1: From the "Main Menu" of your iPod, select "Extras."

Step 2: From this menu, navigate down to "Set combination."

Step 3: Enter a code in the next menu. Rotate the click wheel to dial the first number.

You may also tap the "Previous" or "Rewind" button to navigate through the numbers.

Step 4: After choosing a number, press the "Select" button to make it official; you will then see that the next number is highlighted.

Step 5: Continue with this method to enter the following three digits of the combination.

Step 6: Once you have entered the last number, you will be redirected to the Screen Lock menu.

Step 7: Navigate back to the "Main Menu" and select "Extras."

Step 8: In this menu, you will choose "Screen Lock."

Step 9: Highlight and click "Turn Screen Lock On" (if the combination has not been set, you will be prompted to enter it here).

Step 10: Now, select "Lock" to lock the iPod.

Step 11: When the iPod has been locked, it will display an "Enter Code" screen. There is also a "Key" icon at the top right of the screen indicating that the iPod is locked.

NOTE: The combination you set will remain in place until you reset. When creating a new combination, you will be prompted to enter the previous one.

How to Unlock Your iPod

There are two ways to unlock your iPod. First, you can enter the code by using the same method used to set it. If the correct combination is entered, flashing red numbers will display. If the code is entered correctly, the iPod will successfully unlock and redirect you to the most previous used screen.

You can also unlock your iPod by using the iTunes program. This must be done with the computer the iPod was synced on. Simply connect your device to the computer and open iTunes. After disconnecting it, the iPod will no longer be locked.

A worst case scenario may result in the iPod not unlocking after attempting the methods mentioned above. In this case, you may want to restore the factory settings.

NOTE: Extreme caution must be taken when restoring the factory settings of your iPod. This will delete all of your music, videos, and podcasts that have been synced. The iPod will revert to the way it was after first taking it out of the box. Of course, you can rebuild the content.

The Locked iPod

There are a few things you need to know about how your iPod will react when it is locked. If music was playing at the time of the screen lock, it will continue to play. You can also pause and resume songs during the initial locking. You will not be able to adjust volume or navigate through tracks.

When your iPod has been locked, the Enter Code screen will continue to display until you enter the correct code or shut the device down by pressing the "Play/Pause" button. When you power the device back on, the Enter Code screen will appear again.

Adding Screen Lock to Your Main Menu

If the Screen Lock is something you will be using on a regular basis, you may wish to implement it into the Main Menu for easy access. Simply select "Settings," navigate to the "Main Menu," and choose "Screen Lock." Make sure the Screen Lock status is set to "On." When you go back to the Main Menu, the Screen Lock should appear.

Setting the Clock on Your iPod

Instead of relying on the blare of your clock radio to wake you up, you may want to use your iPod's built-in alarm.

Not only can your iPod be used as a more convenient replacement for an alarm clock, it can also be programmed to automatically shut down at your command. Below are the steps to help familiarize you with the clock feature on your iPod:

Step 1: First, select "Extras" in the Main Menu.

Step 2: From there, select "Clock." This will show a date and time that might need to be configured accordingly.

Step 3: To properly configure the time, you must first adjust the date and time zone. From the Main Menu, click "Settings."

Step 4: In this menu, select "Date & Time."

Step 5: Click "Set Time Zone."

Step 6: Navigate through this menu to find and select your time zone.

Step 7: To set the date and time, you need to backtrack to the Main Menu. Scroll down to "Settings."

Step 8: From this menu, choose "Date & Time." The number representing the hour will be highlighted, displaying arrows pointing up and down.

Step 9: Play around with the numbers to change the time, day, month, and year.

iPod Clock Sleep Timer

The Sleep Timer can be programmed to shut down your iPod within 15, 30, 60, 90, or 120 minutes. To configure your preferred Sleep Timer settings, follow these steps:

Step 1: Select "Extras" from the Main Menu.

Step 2: Select "Clock" in the next menu.

Step 3: Now, select "Sleep Timer" and choose the duration. You will notice that when the sleep timer has been set, a "Clock" icon will display, indicating the remaining time until the "Now Playing" screen resumes.

NOTE: When the iPod shuts down, the Sleep Timer will automatically revert to "Off." The sleep timer must be set each time you choose to use it.

iPod Alarm Clock

Seldom do we use our alarm clocks for anything other than looking at the time or waking up. To use your iPod as a replacement, follow the steps below:

Step 1: Select "Extras" from the Main Menu.

Step 2: Click "Clock" from the next menu.

Step 3: Now, select "Alarm" and turn the function "On."

Step 4: Navigate through the menu and select "Time."

Step 5: Use the click wheel to select the time in the form of hours, minutes, and AM or PM.

Step 6: Next, scroll to "Sound" and select the alarm type.

Wake Up to Your Music

Waking up to the standard alarm is now a thing of the past. An iPod allows you to set your alarm to songs that have been synced to it. Simply select a song or playlist from the "Sound" menu. Wake up to the delicate sounds of Pink Floyd or hop out of bed to the heavy string sounds of Led Zeppelin.

NOTE: If you choose the traditional beep, it will be played through your iPod's internal speaker. If you want to wake up to a song or playlist, your iPod must be connected to a set of speakers or headphones in order for you to hear the alarm.

Voice Memos

The voice memo feature on your iPod is what you will use to record podcasts and other forms of audio. For this to work, you need the aid of a microphone that is compatible with the device.

We will discuss the types of microphones available later in the following chapter that covers iPod accessories.

All models, with the exception of the iPod Nano, iPod Shuffle, and iPod Mini, have the voice memo capability built within. After gathering the required accessories, here is how to record a voice memo on your iPod:

Start by connecting the microphone to your iPod. For the iPod Video, the microphone will be connected to the dock connector. In older iPod models, the microphone will be connected to headphone port.

Upon connecting the device, the "Voice Memo" screen should display.

The iPod Video allows you to select a recording quality. Choose from "High Quality" or "Low Quality." Low Quality will save space on the drive of your iPod, while High Quality will provide improved sound results.

Select "Record" to begin the capture of your audio.

Place the microphone a couple of inches away from your mouth and speak clearly into it.

You can pause or resume the recording by pressing the "Play/Pause" button on your iPod.

When the recording is done, select "Stop" and then "Save." The voice memo will be saved and categorized by a date and time. To hear what you sound like on the iPod, navigate to "Voice Memos" from the "Extras" screen, then scroll through the menu and select the designated recording.

NOTE: The only way to access the Voice Memos menu is by connecting a microphone.

How to Transfer Voice Memos to Your Computer

There are two ways to transfer Voice Memos from the iPod to your computer: manually or by syncing the digital data. When recorded on the device, your voice memos are saved in WAV format and stored into a "Recordings" folder. When the iPod is set to be used as a hard drive, voice memos can then be manually dragged and dropped into your computer.

When your iPod is connected to the computer and configured to automatically sync your music, it will also import the voice memos that have been recorded and place them in iTunes. This completely removes them from your iPod. You can play these voice memos on your computer by clicking the "Voice Memos" icon on the left side of the iTunes screen.

Setting Your iPod as a Hard Drive

In addition to cranking out your favorite jams and the latest music videos, the iPod has a few more worthy functions. The hard drive of this device is similar to that of a desktop or notebook computer. Some models even have more disk capacity than the average computer, such as later versions of the iPod Classic. This abundant amount of space enables the device to act as a major storage unit.

The configurations for this will differ from a Windows operating system to Mac, but both permit the function. This allows your iPod to store tons of files from the hard drive of your computer and act as a boot up disc. This can be done with all models for the iPod Touch and the iPod Shuffle.

Below are the steps for setting your iPod as a hard drive:

Step 1: First, open the iTunes application.

Step 2: Connect your iPod with the USB cable.

Step 3: Click the "iPod" icon when it displays to the left of the iTunes screen.

Step 4: Click the "Summary" tab.

You can then select either "Enable disk use" or "Manually manage songs and playlists." Both of these options will allow you to use your iPod as a hard drive. Choose "Enable disk use" to allow iTunes to automatically update your iPod.

When the "iPod" icon displays on your desktop screen, you can double click it and then drag and drop files to or from the iPod. Be sure that you properly eject the iPod from iTunes before disconnecting it from the computer — the "Do Not Disconnect" message should serve as a reminder.

Transferring from a Mac to a PC

Apple and Microsoft are sometimes able to agree in the way of compatibility. Your iPod plays a huge role when it comes to transferring data. This device even has the ability to move the huge files from the PC at your job and successfully install them on your Mac computer at home.

To transfer data between operating systems, you will need to do the following two things:

1. Make certain you have backed up all your content on the iPod for safe measure.

2.   Reconfigure the device so the Windows operating system can be stored on it. The iPod will be reformatted in FAT, a file recognized by Windows.

Reformatting the iPod is required when transferring data between operating systems. This is because a Mac comes equipped with the ability to translate all of the iPod's files, while a PC must be configured for the transfer.

Storing Secure Data on Your iPod

By using Apple's disk image, you can create an encryption with password protection for your data. The disk image acts a virtual hard drive, its content placed into a typical file. Double clicking the icon will unload the contents of the hard drive onto the computer for your use. When you are done, disconnect the device and the sensitive data is back in your pocket, not on the computer.

Creating iPod Disk Images

If you are operating the Mac X system, then you have all the necessary components to create disk image. Just follow the directions below:

1.   Connect your iPod to the Mac.

2.   Open the "Disk Utility" from the desktop screen.

3.   In this menu, select "New Image."

4.   In the next screen, assign a name to your disk image in the "Save As" prompt.

5.   Select "your iPod" from the list as the source of the save.

6.   Now, choose a size for the disk image in the "Size" option box.

7.   Next, choose an encryption from the appropriate drop-down box.

8.   Make sure the "Format" drop-down box is enabled to "Read/Write Disk Image."

9.   Click "Create."

10. Enter your new password in the following screen.

The disk utility will automatically unload the new data and you can begin to store files to the disk image.

Your iPod to the Rescue

Over time, your computer is bound to endure a few performance issues. Out of the box it runs like a champ then slows down with usage. When this happens, it is normally due to corrupt files stored in the computer's registry. In this instance, your computer may need to be rebooted to repair the primary hard drive and ultimately save the system. Believe it or not, an iPod is the perfect tool to make the rescue.

With a little knowledge, a limited version of your Mac operating system can be installed on an iPod — this includes the disk utility required to reboot the computer. Follow these directions to get started:

1.   Place the disc for your Mac operating system into the appropriate drive.

2.   When the installer displays on the screen, follow the instructions.

3.   Select your iPod volume in the "Destination" area.

4.   Click the "Continue" tab.

5.   Click the "Customize" tab when it appears on the screen.

6.   Remove as many of the Mac's additional components as you can, such as drivers and other irrelevant content.

7.   When the installation is done, disconnect the iPod.

8.   Reboot the computer and make sure it boots from the primary hard drive.

9.   In the case of a computer emergency, connect your iPod and press down the "Options" button as it reboots. You will then see a list of devices capable of performing the reboot.

10.   Select the "iPod" icon and click the "right arrow" to proceed with the boot.

Case Study: Matt Jansen

Matt Jansen

Freelance Writer

Tech.Blorge – Senior Writer

I started writing where many people do, in school, and have been actively engaged in putting thoughts to paper for about 14 years. There is something exciting about writing. It is a great way to create meaning and share it with other people. My hope is that when people read my work, they take away something new or think about something with added perspective.

Case Study: Matt Jansen

Technology by nature is enabling. Writing itself is a foundation technology that we have relied on for thousands of years, and when a new product or service is created that truly adds value and builds on existing technology, I want people to benefit from that. So, I choose to write about it.

TECH.BLORGE.com is focused on providing accurate information, often with a comic twist :-}. My job is to monitor the pulse of technology on the inter-Web and write articles that communicate fresh developments with a unique voice.

In addition to posting at **TECH.BLORGE.com**, I work as a full time Web master at a large company in Michigan; I prowl for human intelligence on the Web over **http:// metaViper.com**, and I have written a fantasy novel now in its third revision.

It took me a while to accept the iPod. I always viewed it as an over-hyped product with a slick form factor that would make me blend in with everyone else. I started with Creative's Nomad II MP3 player and that worked well, though it only stored 512MB worth of songs. A year later I started hitting the gym several days per week and realized how ubiquitous the iPod had become. Seeing the outpour of 3rd party gadgets and accessories I decided to go with the flow.

Because I was at the gym a lot, it was important to have something small that would slip into my pocket. At that point the iPod Mini was declining popularity and everyone was wild about the first generation iPod Nano. It fit my needs, so I bought one along with a lime green arm band.

As a side note, the arm band did not work at all because it would unfasten in the process of doing reps with my biceps.

For a while I listened to podcasts a lot. "TWIT," "Buzz Out Loud," "Mr. Manners," "Grammar Girl," and "NPR" were all in my list of regular updates. Shortly after that interest started I realized I was only interested in listening to the most recent broadcasts. Smart playlists were an important part of making that possible because I could customize by play count, date, or a number of additional attributes.

I also appreciate the click wheel interface and have not seen another MP3 player that matches its ease of use.

At my day job, I plug the ear-buds in and cruise through programming like crazy while ignoring the conversations buzzing around me in other cubes. At least, that works when it is not a day filled with meetings.

How many songs do I have in iTunes? Not many. Just under 1,000. It is enough to keep me entertained and I probably download a couple of new songs every month. Never

Case Study: Matt Jansen

through iTunes, though, as I hate files with DRM and there are cheaper options out there. My favorite right now is **Amazon.com/mp3**.

The artists that currently get the most play on my iPod are Ingrid Michaelson, Timbaland, Lily Allen, and Scissor Sisters.

I would like iPods to equalize volume between MP3s automatically because it is aggravating when my ears are blasted because I advance a track. Here are some other features I would like to see in future iPods:

— FM Tuner

— Better quality earbuds included

— Battery that is easy to replace

— Biodegradable materials used in manufacturing process

Accessories for Your iPod

A new iPod comes equipped with everything you need to get started. This includes a set of ear-buds, headphones, and a charger. While the basics may be fine for playing music, there are many accessories available to enhance the experience with your iPod. Some were designed specifically for the device and others are merely compatible. All will make your iPod a bit more safe or enjoyable.

Official iPod Accessories

iPod Nano Armband

This armband enables your iPod Nano to be the perfect workout companion. Its lightweight design allows you to place the device in your pocket or secure it around your wrist or arm and snap it shut with a set of adjustable fasteners. The iPod Nano gives you skip-resistant playback, allowing you to pump iron or run at Olympic speed on the treadmill.

This accessory is specifically designed for the iPod Nano, third-generation model. The iPod Nano armband comes in grey.

Nike+ iPod Sport Kit

The iPod will act as your own personal trainer by using the Nike+ Sport Kit. Exclusively designed for the Nike+ sports shoe and the iPod Nano, the built-in wireless sensor and receiver provides the feedback from your workout in real-time. These statistics can then be viewed from your Mac or PC. Simply insert the sensor into the custom-made pocket underneath the sole of the shoe and connect the receiver to your iPod Nano's dock connector. When you walk or run, the sensor will send data to the iPod with details of your pace, distance, time, and the amount of calories burned.

From there, you can sync the iPod to transfer data of your workout to the iTunes program or the Nike+ Web site. This allows you to monitor your performance for the day, re-evaluate your goals, or even challenge others to a virtual workout.

XtremeMac InCharge Auto Charger

This accessory enables you to charge both your iPod and iPhone when you are on the go. Backed by a lifetime guarantee, the XtremeMac Incharge Auto Charger can be used with any standard 12-volt outlet.

This charger also has a built-in, self-resetting fuse. The detachable cable comes in just the right length, allowing you to connect your iPod or iPhone to any vehicle. This package includes the XtremeMac InCharge Auto, a power plug, and a USB cable compatible for both the iPod and iPhone.

Apple iPod Earphones

Much more comfortable than the set of ear-buds that come with your device, the Apple iPod Earphones are idea for a vigorous workout at the gym or study time at the computer desk.

These advanced earphones are available for all iPod models.

Belkin TuneBase FM with ClearScan

This handy accessory allows you to amplify the power of the iPod through your car stereo system. It is programmed to automatically pick up the clearest FM radio signals to reduce static. TuneBase FM features include:

- ♫ Compatible power adapter to charge your iPod

- ♫ Special PRO settings to configure audio and adjust volume

- ♫ Flexible neck with integrated antenna

- ♫ Built-in line-out port for integration with car stereo input or cassette adapters

In this package, you will receive Belkin TuneBase FM, along with several support cradles to match up with the many different iPod models.

iPod Touch Incase Protective Cover

Get full protection for iPod touch without limitation. The Incase Protective Cover will not cramp your style as it gives easy access to the iPod's interface. Its lightweight, form-fitting structure makes it easy to transport.

The Incase Protective Cover makes it simple to work the touch screen and it even allows for attachment to your dock connector so it can be charged when it is the case.

iHome iH6 Dual-Alarm Clock

This accessory can be used after mastering your iPod's alarm clock. The iHome iH6 Dual-Alarm Clock Radio is perfect for setting a wakeup call to music. It comes with a multifunctional remote control, which is compatible with the radio and your iPod. The finely tuned Reson8 speaker will pump natural sound, regardless of where you are in the house.

With all of its great features, this may be one of the most useful accessories for your iPod.

The Alarm Features

♪ Allows you to wake up to the best music on your iPod, an AM or FM radio station, or the traditional "buzzer"

♪ Automatic reset function automatically resets alarm for the following day

♪ Alarm battery is backed up to maintain time settings and wake time in case of electrical issues

♪ Rise and shine in peace with the Gentle Awake alarm sound

♪ Alarm backup for iPod when the device is designated as the "wake-to source" (when the iPod is not connected to the dock, you can still be awakened by the alarm clock)

Radio Features

♪ Crisp and clear signals for AM/FM

♪ 12 AM/FM preset stations

♪ Advanced Automatic Frequency Control system

♫ External antenna to get the best signals for AM radio

♫ Built-in equalizer and balance controls to perfect your audio output

iPod Features

♫ Dock inserts for comfortable compatibility with all dock- accessible iPod models

♫ Built-in port for charging and using the iPod shuffle model

♫ Capability to charge your iPod shuffle and different iPod model at the same time

♫ Patch cord for connecting non-dock accessible iPods.

Auto-Set Features

♫ Pre-installed factory settings for time and date

♫ Automatic adjustments to seven different time zones with the simple press of the a button

♫ Automatic clock adjustments for daylight-savings time schedule

♫ Reson8 Sound Chamber

♫ High quality amplifier created for reproducing clear and accurate audio sound

♫ Specially designed speakers to produce clearer high notes, more detectable midrange, and deeper bass

♫ Chamber crafted to reduce vibration and static sound

♫ Custom stereo settings that adjust according to your music genres

The display and design of this accessory is immaculate, featuring two high-fidelity speakers entrapped in isolated chambers. An extra large LCD display screen makes for easy viewing and comes with adjustable backlighting. The time, alarm info, iPod contents, radio tuning, and volume level can all be displayed on the LCD.

In the Package

♫ iHome iH6 unit

♫ Remote control

♫ Docking inserts

♫ Integratable audio cable for stereo analog

♫ AM antenna

♫ AC power adapter

♫ A pair of AA batteries for backup power supply

♫ Apple Universal Dock

Not only can you charge your iPod with this accessory, you can connect to a TV or set of speakers as well. The Apple Universal can be controlled by remote control, giving you the power to command your devices from across the room.

Apple Universal Dock Connector

The Apple Universal Dock Connector is a stylish unit that can be used for charging your iPod, syncing, and so much more. The interchangeable dock adapters make it compatible with the Apple iPhone and any of the dock accessible iPods.

Detailed Features

♫ Compatible with the USB cables of both your iPod and iPhone

♫ Able to be connected with a standard electrical outlet by way of USB Power Adapter

♫ Able to be connected to a stereo system or separate speakers by using audio cable to play music from your iPod

♫ Enjoy digital photos, slideshows, and video content on a TV or alternate video device with compatible AV Cable

In The Package

♫ Apple Universal Dock

♫ Remote Control

♫ Five Dock Adapters to integrate with the iPod Classic, iPod Nano, iPod Video, iPod Touch, and the iPhone.

Belkin Leather Folio

This stylish case was specifically designed to protect your iPod Classic. The sturdy leather saves your device from the intense wear and tear of everyday use and a lifetime warranty gives you the comfort of knowing your iPod Classic is always covered.

Features

♫ Soft and sturdy leather for durability

♫ Transparent protective screen on case for easy navigating

♫ Velcro fastening straps for additional support and security

♫ Ability to access to the dock connector and hold switch

♫ Compatible with all iPod Classic models

Belkin Leather Folio (iPod Nano)

Belkin also offers a protective leather case for the iPod Nano. This one is a bit different in design with a special seam that keeps the iPod snug on your person while keeping it free from everyday wear and tear.

Features

♫ Transparent protective screen on case for easy navigating

♫ Magnetic closure for secure fastening

♫ Access to the hold switch

♫ Comfortable suede lining

♫ Access to the headphone jack

♫ Comes in the colors black and chocolate.

Incase Leather Sleeve (iPod Nano)

This protective sleeve features a clear screen for easy navigating. Its form-fitting design gives access to all of your iPod's controls and the dock connector.

Features

- ♫   Comfortable, custom suede lining

- ♫   Belt or strap for secure attachment

- ♫   Custom fit for your iPod Nano

- ♫   Comes in black

Incase Protective Cover (iPod Touch)

This accessory is the perfection solution for securing your iPod Touch. A specially designed topographic pattern adds additional grip while giving your iPod Touch a sense of style.

Features:

- ♫   Complete access to your iPod's controls

- ♫   Access to the dock connector

- ♫   Ability to accept a charge while in the case

- ♫   iPod Radio Remote

Enjoy the best of FM radio, directly from your iPod. Simply plug in the Radio Remote and select the "Radio" option from the iPod's main menu. View the tuning on the screen and dial stations with the click wheel. The iPod Radio Remote allows you to mark your favorite stations for easy access. You can also access radio stations with either your iPod or the remote.

When listening to stations that support the Radio Data System standards, you will be able to view the radio station data or the song and artist on the display screen.

The iPod Radio Remote is very capable, enabling you to control your music, video, and even slideshows. You can also adjust the volume, advance and rewind tracks, and change radio stations and power of the radio from a good distance. This remote can be easily clipped on your pants pocket or shirt collar for easy access.

One of the best things about the iPod Radio Remote is its extended life span. Powered by the iPod, the Radio Remote does not require batteries to operate, assuring you that the device will keep going for a long time.

Power Support Anti-Glare Film Set (iPod Touch)

The annoying glare reflecting from your iPod will no longer be an issue with this accessory. The Power Support Anti-Glare Film Set makes your iPod Touch easier to view while providing reliable protection against the scratches and dust that result from everyday use.

This accessory uses the best film available, featuring a smooth, matte finish specifically designed for your iPod's LCD screen. Its adhesive side is constructed of polymer plastic that will leave no residue when removed.

In The Package

♪ Two separate protective films for the front and back of your iPod.

iPod Microphone Accessories

In the previous chapter, we touched on the iPod's ability to record voice memos. In this section you will learn that some of the best accessories you can find are iPod microphones. With the appropriate device, you can record everyday conversations and interviews, and even make mixing and editing adjustments to your online podcasts.

Capturing an audio interview was once quite a task, requiring you to transport an unattractive portable cassette recorder wherever the captivating story was. Methods of recording have advanced slowly, from microcassette recorders up to the much more portable digital audio recorder. An iPod functions similarly to a digital audio recorder with its disk storage drive.

After figuring iTunes into the picture, one may argue that the iPod is even more advanced due to that very factor.

Most of the iPod microphones you will find on the market allow control over mono or stereo functions, as well as the bit rate and volume. They also permit you to save the recording as an audio file on the hard drive of the device. Some of them consist of one microphone, while others have two built into the unit. You will also find a few with built-in ports that allow the use of external microphones. Descriptions of a few of the most popular microphones for your iPod follow.

iTalk Pro iPod Microphone

Griffin Technology is one of the leading providers in quality iPod accessories. Let us now introduce you to one of their most popular iPod Microphones, the iTalk Pro. This device is simple to use and brings out the power in your iPod's Voice Memo feature. To begin recording, follow these four steps:

Step 1: Remove the iTalk Pro from package and remove the plastic covering for the connector.

Step 2: Next, insert the connector into the iPod headphone port.

Step 3: Give the two devices a few seconds to detect one another, then press the "iTalk" button in the middle of the iTalk Pro device. Proceed to record your voice.

Step 4: Once you are done recording, press the iTalk button.

NOTE: Of course this procedure will be slightly different depending on the brand of microphone. Aside from that, recording with an iPod microphone is pretty easy.

You can find out more information about the The Griffin iTalk Pro at **www.griffintechnology.com.**

XtremeMac MicroMemo

This is certainly one of the most unique iPod microphones you will encounter. The XtremeMac MicroMemo attaches to a dock connector port at the bottom of your iPod. It features a rather flexible microphone unit that is easy to detach. The built-in speaker allows playback of the content that has just been recorded.

You can expand on the power of the MicroMemo by attaching an external microphone to the mini jack connector. Set your recordings to "Low" quality to preserve disk space or "High" to prompt the best capable output.

There are different versions of the XtremeMac MicroMemo available for the iPod Video and the iPod Nano. You can find out more about these iPod microphones at **www.xtrememac.com**.

Belkin TuneTalk Stereo

When you are looking for exceptional quality and performance in an iPod microphone, you do not have to look any farther than the Belkin TuneTalk Stereo. It is compatible with the iPod Video and attaches to the dock connector.

This unit contains two miniature microphones that are positioned in the center. When the recording gets rough, you can connect an additional microphone into the mini jack connector.

The recordings can be adjusted in real time by rotating the click wheel on your iPod. Keep track of the editing process by viewing the clipping editor on the screen.

You can find out more information on this iPod microphone at **http:// www.belkin.com**.

Belkin Voice Recorder

This microphone is suited for older iPod models. Its operation as an add-on is universal, as it connects to the iPods headphone port. This is strictly a mono tone unit, but does come equipped with a built-in speaker that allows the playback of your recordings.

You can find out more about this iPod microphone and its compatibility at **www.belkin.com**.

NOTE: Your recording ventures may one day excel to telephone interviews. Any iPod voice recording device that comes with an input for an external microphone is also capable of being connected to most telephones.

Unofficial iPod Accessories

22Moo Introduces SeepuStar iPod-Compatible Video Glasses

Here is one of the most unique iPod accessories you will find on the market. 22Moo will send your iPod soaring into the future. The SeepuStar DV230 is a portable video display that can be worn over your face like a pair of glasses. It has been designed to display a 35-inch static-free virtual screen with enhanced quality. These video glasses are compatible with almost any video source including a DVD Player, PS2, Xbox, and even your iPod video. The display is easily adjustable, allowing you to view video from the most comfortable angles. It includes adjustable arms to fit a larger sized head and can also be adjusted to comfortably fit an individual who wears prescription eye glasses.

Product Details

♫ 35-inch virtual display

♫ Full color 320 by 240 resolution

♫ Lithium powered battery — up to six hours of video playback

♫ Built-in stereo speakers for quality audio

In the Package

♫ Lithium battery

♫ AC power adapter

♫ AV cables

♫ Nose bridges

♫ Charger (USB cable)

♫ Instruction manual

iPod-Compatible Pet Carrier

Here is another accessory that seemed to come out of left field. With that said, I think it is safe to say that a few people can make use of this one — the Lifepop Stereo Pet Carrier. Now all you pet lovers out there can travel with your pooch in style as they jam your latest playlist.

This unique concept features a soft padded bed and built-in speakers. Your iPod attaches easy to the mini-plug. The top of the carrier even opens to give your pet a bit of fresh air as they rock on.

iPod Compatible Bed

This next creations was brought to your by New Zealand based company, Design Mobel. They were the first to develop and release an iPod-compatible bed, ironically named "Pause."

This comfortable bed comes equipped with a docking station for your iPod and a Bose speaker system. Some of these beds even come with shelves that are intended for additional audio equipment.

An iPod-compatible bed is not so far fetched — this item has actually become popular to those who can afford it. It is the ideal toy for people who cannot fall asleep without music.

The iKitty

Here is one of the most adorable iPod accessories you will find on the market. The ingenious developers over at Speck Products created the iKitty just for your iPod Nano. This cute toy is also a protective case for the device. Your iPod Nano easily slides into the iKitty's belly — this makes it stand on four legs resembling a real-life cat. The iKitty is available in black and white.

iPod Gloves

Next, we have the iGlove Multi, created by Marmot. These are iPod-compatible gloves that do a decent job warming your hands while allowing you to access your tunes. The company says they can be worn by themselves, but they work as perfect liners for heavier gloves.

The iGlove Multi is designed for great flexibility, allowing you to move your thumb and fingertips with ease over the click wheel. The body of the glove consists of silicon printing to increase your grip and durability.

I like the concept here — it is a good way to enjoy your iPod regardless of the frosty temperatures. I think that a glove capable of comfortably storing an iPod would be an even better invention.

Neon iPod Headset Cables

Here is another cool accessory that adds some fun to your jam session. These iPod-compatible cables are futuristic neon and attach to ear-buds or a headset. The neat part is they flash at the beat of your music. These neon cables are powered by any brand of "AAA" batteries.

The neon headset cables were originally created in Japan but can be found on a few unofficial iPod accessory Web sites.

The iPillow

Here is the perfect companion for your iPod bed — the iPillow from BrookStone. This comfy device is compatible with iPods and virtually any other brand of MP3 player on the market.

The iPillow has built-in speakers powered by six "D" batteries. I am sure it can pack a punch.

Your iPod fits snugly in the center of this device. I would not suggest starting any pillow fights with this device as the pocket does not provide much padding. Aside from that, this is another clever concept that just might help you rest a little better.

The iPod Mouse

I found this to be one of the more intriguing accessories. The iPort, by Sonance, is the perfect stereo component to complete your home entertainment system. This is an in-wall system that integrates smoothly with all iPod models. The system is compatible with all Sonance home audio products and other stereo systems as well.

This system comes with a docking station to charge your iPod while giving you complete access to all the device's controls. The iPort is up there in price but comes with all the accessories and cables to work with any iPod.

iTrip

Here is another dynamic iPod-compatible product brought to you by Griffin. This is one accessory that really compliments your iPod in the way of presentation and functionality. The iTrip comes with upgrades that stand out amongst previous FM transmitters. This device outputs quality performance yet is very easy to operate.

The iTrip is equipped with a large LCD screen, brightened by a backlight. The easy-accessible knob on the side of it enables you to change frequencies in a heartbeat — simply turn to a preferred frequency, then click the knob and set it. The iTrip will automatically store your settings for future use.

What makes this device so unique is the advanced broadcasting functions. The iTrip features LX and DX modes that allow you to make adjustments for the best audio playback in any environment. The DX mode produces a low-level of sound, which gives you clear audio in crowded places with lots of noise in the background.

The iTrip is able to set frequencies in U.S. and International modes. This means your options will be filled with relative stations, as opposed to surfing through tons that do not fit your criteria.

This accessory is sure to bring you the best in FM radio with its volume enhancement control. This handy function saves you the hassle of playing with your buttons as it automatically adjusts to filter out audio distortion. If the volume on your iPod is too high, the iTrip will detect and adjust it accordingly. It will lower the volume to a reasonable level that produces crisp, clear sound.

Features

♫ Large LCD screen for easy viewing

♫ Seamless housing for attaching your iPod

♫ No batteries required — runs off charge of the iPod

The Griffin iTrip is available for the iPod Photo, iPod Classic, and the iPod U2 Special Edition model.

iVoice

This handy accessory proves that the iPod easily has the potential to be the most expandable entertainment system in your home. The iVoice by Athena is a complete docking station and sound system for your iPod. This extraordinary device was built to give maximum performance, and it is able to play louder, longer, and with much better quality than many other brands.

Product Features

♫ Allows video output for viewing content on your iPod Video, iPod Photo, or TV

♫ Audio input enables the system to work like a PC-based speaker

♪ Includes wireless remote for controlling your music from far away

♪ High-gloss finished cabinet is able to be mounted on your wall

♪ Charges and plays iPod while docked

Case Study: Michel Moore

Michel Moore

Essence Best Selling Author

Hood Book Headquarters, Owner

I have been a book distributor for a little shy of two years now. Hood Book Headquarters is an African American owned business that specializes in urban fiction and street literature. A regular day for me consists of driving to various locations picking up new book titles.

Growing up in the urban setting of Detroit, Michigan, I used writing as a form of escape from the blight of the city. At nine or 10 years old, I began developing short stories. My road from being self-published to an Essence Best Selling author can be described in one, no, two words – "Hard Work!" I saved my money, printed up a few cases of books, and hit the streets of New York.

Out and about selling my book, I needed entertainment and something to keep me going — this is when I turned to an iPod. The best feature is the ability to download music and take it on the road with you.

iPods are cool, but with all the movies that are available I sometimes wish mine had an unlimited amount of gigabytes.

I am not too familiar with any other brand of MP3 player. I have heard of the Creative Zen and the Zune, but as far as I know they are not any serious competition to the iPod.

Everyone who loves music should enhance their experience with an iPod. If I had to recommend one model it would be the iPod Video.

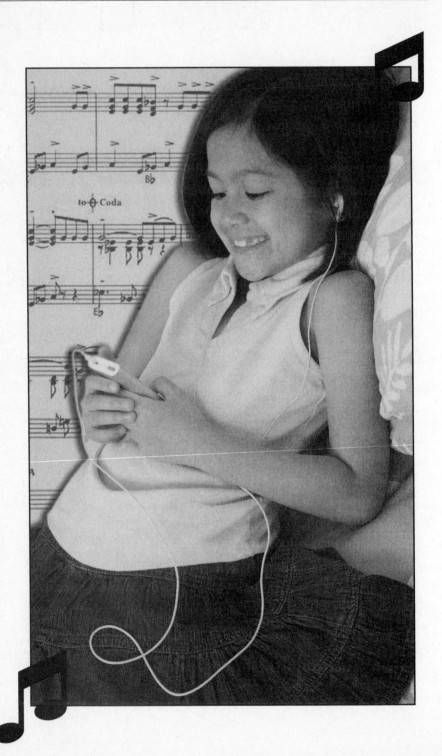

Other MP3 Options

The year of 2005 proved that portable multimedia is a competitive market. Apple can take much of the credit, considering how they shook up the industry with iTunes and the iPod. As other companies continued to develop products to compete, industry leaders worked hard to agree on MP3 standards. One of the biggest issues was compatibility with outside accessories. While most MP3 players contained a dock connector, many could not easily be attached to a car radio or home audio system.

Microsoft was one company that supported the move to unify portable multimedia device developers into making units that were more compatible. Meanwhile, Apple continued to thrive. Their iPods were dock-accessible and able to connect to exterior speakers, car stereos, and several other devices. On the other hand, multimedia companies, such as Creative Technology, had their own brand of dock ports. This put them at a disadvantage as many manufacturers of accessories were reluctant to develop new editions to work with such a large amount of formats.

Around this time, most multimedia accessories were strictly designed for iPods as Apple towered over competitors. These were profitable times for both Apple and manufacturers of their add-ons. It has truly been a struggle for some, but strides have been made. As it stands, the iPod is still the most recognizable device in the industry. With that said, there are a few other brands that are well worth mentioning.

When looking to compare other brands of MP3 players to the iPod you should survey three areas: capability, compatibility, and value.

>   **Capability:** By now you have learned what an extraordinary device the iPod is. At the same time, there are a few other brands that have it beat in the way of capability. One example is that most iPods do not come with a built-in FM radio feature.

>   **Compatibility:** This may be another issue for many reasons. Several companies have caught up to industry standards but a few of them still are not able to support as many external devices as the iPod. However, there are newly released MP3 players that have taken a jump on the iPod by setting new standards of compatibility.

>   **Value:** The value of an MP3 player can be measured in cost and personal benefit. As simple as they are to work, iPods are just too advanced for some. Perhaps you do not need all the fancy features. The iPod enthusiast may think otherwise, deeming those features priceless.

Regardless of the factors mentioned above, any portable digital music player that hits the market will have to be measured against the iPod. Apple set the bar high and continues to improve with each new model.

To give you a better idea of what actually is on the market, we have combined detailed information on a few notable mentions, as well as the multimedia players that can be considered rivals of the iPod.

Rhapsody MP3 Players

RealNetworks' music service became such a hit that Rhapsody received its own line of digital media players. Here is list of the popular devices they currently have on the market:

Sansa Rhapsody

RealNetworks gives users the ultimate music experience with its Sansa Rhapsody. This device plays MP3 formatted music, video clips, and also supports photos.

The Sansa digital audio player is the first to offer Rhapsody Channels — non-stop mixes automatically categorized based on your preferred music genres and listening history.

Features

- ♬ Pre-loaded with hundreds of songs through Rhapsody Channels and your playlists

- ♬ Video playback

- ♬ Built-in digital FM radio tuner

- ♬ Voice recording integrated with built-in microphone

- ♬ The Sansa Rhapsody is very portable with a sleek and thin design. Its casing is constructed of a sturdy alloy metal that makes it scratch-resistant. A large 1.8-inch screen makes for easy viewing. This device is powered by a replaceable, rechargeable lithium-ion battery that gives you up to 20 hours of audio and video playback.

♫   The Sansa Rhapsody comes in three versions offering two, four, and eight gigabytes on a flash-based storage drive.

clix Rhapsody

One of the best attributes of this model is its attractive design. The clix Rhapsody is perfect for viewing your video content with a 2.2-inch color display screen. RealNetworks' exclusive D-click navigational system makes this device one of the most portable MP3 players on the market.

Features

♫   Supports MP3 formatted music, video clips, and photos

♫   New music continuously provided through Rhapsody Channels

♫   Built-in digital FM radio tuner

♫   Voice recording integrated with built-in microphone

♫   The clix Rhapsody is powered by a lithium-ion battery that gives you up to 24 hours of audio and video playback. RealPlayer offers you two versions of this model, the iRiver clix Rhapsody with four gigabytes, and the iRiver clix Rhapsody with eight gigabytes.

Sansa View

This is the most sophisticated Rhapsody player of them all, a device that expanded on RealNetworks' video capability. The Sansa View pleasantly blends full-motion video with all the features of a powerful MP3 player. A vivid 2.4-inch widescreen makes for easy viewing of your favorite movies and TV shows while on the go.

Features

- ♪ Supports music, videos, photos, online radio, and audiobooks

- ♪ New music continuously provided through Rhapsody Channels

- ♪ Built-in FM radio tuner with 20 digital preset frequencies

- ♪ Built-microphone for recording voice memos, and FM radio

- ♪ Powered by a rechargeable lithium polymer battery — up to 35 hours for audio playback and up to 7 hours of video playback

- ♪ The capacity of the Sansa View is flash-based, offering eight and 16 gigabytes of storage

All RealNetworks portable multimedia players required Rhapsody's To-Go service plan. This gives users unlimited access to millions of songs that can be easily imported into any Rhapsody device. Rhapsody's To-Go service is currently limited to users in the United States.

NOTE: All Rhapsody portable players are PC-based at this time.

Sansa Clip

While it has not spent much time in the spotlight, the Sansa Clip is a very remarkable unit — sleek and sophisticated, small and functional. This MP3 player was created by the company SanDisk, offering devices with one or two gigabytes of storage capacity — about enough space for 250 to 500 songs.

The feature of the Sansa Clip is certainly the price. The one gigabyte is only $40, while the larger two gigabyte version is just $60. This device comes in black, blue, red, and pink. It easily clips to your belt or fits perfectly in a small purse.

The Sansa Clip has been frequently compared to the iPod Shuffle (in which it has the Apple model beat in the way of price). It also includes an FM tuner, built-in voice recorder and a subscription link for purchasing online music. Navigating with the device is simple as all of your artists, song titles, and albums can be accessed from the display. This is another advantage over the iPod Shuffle since it has no screen.

Considering the extra perks of having access to the FM radio and a voice recorder, this is a smart choice — especially for the price. The Sansa Clip may be ideal for someone looking to buy his or her first MP3 player.

BeoSound 6

This next MP3 player does not match up so well with the iPod. You will be left scratching your head after comparing the features of this device to its price. For starters, the BeoSound 6 does not play video or FM radio. This player continues to amaze as the price tallies in at around $600 which only gives you four gigabytes of storage space. This price is nearly three times as much as the eight gigabyte iPod Nano's price.

This device does produce quality audio output as the songs are able to be imported at a higher bit rate. This means that files are much less compressed which makes for more clear and natural sound.

The most attractive feature about this player is a set of comfortable earphones that comes included with the package. They are quite unique in the way they wrap behind ear. The buds can be adjusted up or down to assure a snug fit. It also comes with a stylish leather protective case.

This device does have exceptional sound and solid presentation. As far as its capability as an MP3 player, the BeoSound 6 fails in comparison to any other brand that will be mentioned here. The asking price is just a bit steep considering how limited it is.

Creative Technology Versus Apple

Creative Technology is one company that has made some headlines in the multimedia player industry. Some would even go as far to say that they have achieved a victory over Apple. This praise came after they were officially awarded the patent for MP3-based technology. In a sense, their devices were the face of the industry as opposed to iPods.

Creative Technology are the developers of Nomad Jukebox MP3 player and its offspring, the Creative Zen. The company was proud of this acclamation, taking credit for being the first brand of MP3 player before the iPod was even thought of.

The battle of MP3 title between Apple and Creative Technology was one that stretched out to a three-year dispute. After several attempts of filing for the patent, Apple was continuously rejected and beat out by Creative Technology. The ironic part of it all is that the iPod remains ahead of the game to this date.

Determining which company should be credited with popularizing MP3 players is certainly an arguable topic. One fact that cannot be debated is that Apple is the first to capitalize on the market in ways of advertising and branding. Their ability to continuously satisfy and improve upon public demand is what has set them far apart from the competition.

The NOMAD was the first brand of digital audio players by Creative Technology. This MP3 device was composed of two distinctive models — THE NOMAD and THE NOMAD Jukebox. The basic NOMAD was a player that utilized flash memory, while The NOMAD Jukebox utilized a microdrive for storage space.

The first NOMAD device was actually a rebranded model from Samsung Electronics — the YP-D40 player. Certainly unique at the time, the devices had a capacity of 64 megabytes, which we know by now is not much at all. These players worked well as MP3 storage devices, allowing them to be connected to an operating system, then accessed and used like other removable media such as CDs and floppy disks.

Both of these Creative Technology devices were discontinued in 2004. The original NOMAD evolved into the NOMAD MuVo while the NOMAD Jukebox took on the Zen product line of MP3 players.

Creative Zen

After putting the NOMAD player to rest, Creative Technology resurfaced with the Zen. This device put credibility back into the Creative name with a device that actually seemed to match up to the Apple super power. This player is actually the predominant device in many Asian markets, with a strong show of support in Singapore, the base location for company headquarters. Creative Technology reclaimed some of the limelight in the United States as their product gained favorable recognition. Three models from the Zen line have received awards in their respective categories for the best of CES (Consumer Electronics Show) from 2004 to 2006.

All Zen models are highly compatible with MP3 and WMA formats, an attribute that instantly gave it a slight advantage over the iPod. Many versions also support WAV files and audible formats to play audiobooks. Zen models come included with drivers and the Creative MediaSource, Creative Technology's version of iTunes. This application works as a media database for transferring and syncs content exclusively for the Zen MP3 player.

Creative Technology has tagged these devices as PlaysForSure certified, meaning they are compatible with almost any file format by way of Media Transfer Protocol. The latest models of Zen have been designed to exclusively support the Microsoft Windows XP and Vista operating systems for PC computers.

Popular Zen Models

The Zen V Plus

The first flash-based portable digital medial player by Creative Technology was released on June 20, 2006. The Zen V Plus featured a scratch-resistant display — a 1.5-inch OLED with 128 by 128 resolution. This model is capable of exceptional video playback and is equipped with an FM radio tuner, the first of the Creative Technology players with this capability. The Zen V Plus has a built-in microphone for easy voice recording, along with special wires for line-in recording purposes. It is powered by a lithium-ion battery. This rechargeable unit lasts up to 15 hours for audio playback.

The ZEN V Plus was the winner of the CNET Editor's Choice Award during the same month it was released. The one gigabyte version of this model comes in black and white with orange trim and green trim. The two gigabyte version comes in black with blue trim.

The Zen V Plus received a notable upgrade on September 29, 2006. These revisions applied to the blue and black model as the storage space increased to eight gigabytes. Creative Technology became more popular after a successful merger with the national Breast Cancer Foundation. At this time, they released a pink version of the two gigabyte version. The Zen V Plus also supported the cause by embedding the National Breast Cancer Foundations symbolic ribbon on the device.

Zen Neeon2

Released on August 2, 2006, this player is the successor to the Zen Neeon. The Zen Neeon2 features a 1.5-inch CSTN-LCD and displays a resolution of 128 by 128. This model is compatible with MP3, WMA, and WAV file formats. It also displays vivid JPEG images and plays AVI formatted videos. Files are transcoded upon upload procedure. The Zen Neeon2 comes with an FM radio tuner, built-in microphone, and line-in recording capability in WMA format.

The rechargeable lithium-ion battery gives you up to 20 continuous hours of audio playback — up to 8 hours for video playback, the largest amount of time for any Creative Technology portable multimedia player.

One gigabyte models of the Zen Neeon2 are available with a black, silver, blue, pink, or orange backplate and a black piano design covers the front. Four gigabyte models are only available in black and silver.

Zen Vision W

Creative Technology continued to improve on their MP3 players. The company released the Zen Vision W on September 17, 2006. This portable media player featured a 4.3-inch widescreen TFT-LCD display. Like its predecessors, this model includes a built-in microphone for easy voice recording along with FM radio capability. The Zen Vision W supports MP3, WMA, and various file formats. The interface of this device is similar to previous models and has the ability to sync contact information from the Microsoft Outlook application.

The Zen Vision W comes in black, with storage sizes from 30 to 60 gigabytes. This device also comes equipped with a CompactFlash reader to increase file capacity and manage the direct uploading of images.

Zen Stone

Creative Technologies released the Zen Stone on May 3, 2007. This model posed as the perfect rival to Apple's iPod shuffle in the way of design. Very small, compact, and screenless, the Zen Stone is a one gigabyte unit that is compatible with MP3, WMA, and audible file formats.

The Zen Stone comes in black, white, blue, green, red, and pink. This device is reasonably priced at $39, a bit cheaper than the iPod Shuffle.

Zen Wav

The Zen Wav became available on May 25, 2007, exclusively in Singapore. The storage drive of this MP3 player is flash-based; the features are slightly more advanced than previous models. This unit supports MP3, WMA, and WAV file formats, and includes built-in stereo speakers for enhanced audio input. This is the first portable multimedia player by Creative Technology to feature a 16-bit color LCD display, up to 30 hours of continuous audio playback (20 hours when stereo speakers are enabled), the ability to store and read eBooks, and displays three different times zones.

The Zen Wav comes with storage capacities of two and four gigabytes. This unit became available for U.S. customers in August of 2007.

Zen Stone Plus

The Zen Stone Plus was released domestically on June 29, 2007. While slightly larger than the original Zen Stone, this model is identical to it in the way of its interface design. The Zen Stone Plus comes with a monochromic OLED display screen, a built-in FM radio tuner, and a stopwatch — the first portable player by Creative Technology to house this feature.

The lithium-ion battery gives you an estimated 9.5 hours of audio playback and has capacity of two gigabytes. Just like its predecessor, the Zen Stone Plus comes in black, white, blue, green, red, and pink. The Zen Stone Plus was released internationally on July 26, 2007.

Creative Records

Creative Technology reached a milestone of their own on September 14,

2007. Upgrades to the Zen V Plus model were flash-based with storage capacities of two, four, six, and eight gigabytes. Their new model holds 16 gigabytes — a record number in the industry amongst all flash-based portable digital medial players.

This version of the Zen handles easy at 0.44-inches thick — certainly the slimmest of the brand so far. This is also the first device from the Creative family to come with an SD card slot to support unprotected AAC files. It also gives a great look at your playlist by way of the 24-bit display screen.

Creative Technology has been around for sometime now. Though Apple improved on the original concept, products such as the Zen are becoming more comparable. If anyone is able to keep pace in the fast paced, multimedia industry, it would have to be a company that was there from the beginning.

Apple Versus Creative Technology — Conclusion

Creative Technology was finally granted the patent on August 25, 2005. The company first applied on January 5, 2001 and Apple followed up with their request on October 22, 2002. Creative Technology immediately stated that it would actively pursue respect of its MP3 patents. The company requested that all manufacturers in the industry utilize the same navigational system for reasons of compatibility. Apple was the only company accused of violating these terms.

On May 15, 2006, Creative Technology declared a lawsuit against Apple Inc. for infringement of its patent in regard to the Zen player. They also requested that Apple be investigated by the U.S. Trade Commission for violating other laws. Creative's claim was that Apple may have breached set trade laws by importing iPod devices into the United States from other countries. On the same day, Apple reacted by counter-suing Creative Technology. The company filed two separate lawsuits claiming that Creative had infringed upon their patents.

The controversy between these two multimedia powerhouses has since calmed down. The companies have reached a legal settlement. Apple paid $100 million to compensate those accusations while Creative Technology agreed to join the worldwide "Made for iPod" accessory program.

Archos Media Players

Archos was established in 1988. This France-based consumer electronics company specializes in manufacturing a wide range of multimedia devices, from typical MP3 players to portable video players (PVPs). Archos designs media players with both USB flash and hard drives. They also have built a reputation for a quality line of digital video recorders and the Portable Media Assistant — a highly advanced media player that also has Wi-Fi wireless Internet capability and PIM software.

While Archos creates a wide range of media products, they also have a variety of players that directly rival the iPod. The Archos 105 has been frequently compared to Apple's iPod Nano. This device is small, yet very durable. It features a one gigabyte flash drive for storage capacity with a 1.8-inch display screen.

The screen on this player is a bit smaller than the iPod Nano's, making it difficult to view video content. The Archos 105 only supports a resolution of 160 by 128 resolution — half of what you get with the iPod Nano. This player also lacks in capability as it only plays WMV formatted files in a 15-frame-per-second viewing rate.

The interface of this device is fairly manageable. It holds the basic controls found on any other MP3 player. Sifting through your music, photo, and video content is easy — the browser mode allows you to get online. What is missing from the feature list is an FM radio tuner, voice recorder, and line-in recorder. While the iPod Nano does not support these features either, several other portable media players do.

Considering their extensive line products, Archos does not necessarily need to compete with the iPod to survive. Having a lineup of comparable MP3 players will only enhance this company's image in the long run.

There is one Archos device that you may want to look for — this player actually matches up well with the iPod. The Archos 605Wi-Fi gives exceptional video playback and comes with large capacities of 30, 80, and 160 gigabytes.

The Death of the Rio Media Player

Diamond Multimedia can arguably be called the first company to invent the MP3 player. The Rio PMP300 was released in September of 1998. This device did quite well that year as sales from the holiday season exceeded all expectations. The release of this device is what ultimately sparked global interest in digital media products.

Diamond Multimedia was met with controversy when the Recording Industry Association of America filed a lawsuit against them. It was alleged that the Rio PMP300 promoted the duplication and distribution of illegal music. Diamond Multimedia inadvertently prevailed in this legal battle on the heels of another case — the Sony Corporation vs. Universal Studios. It was then that digital audio players were ruled to be legal devices, which helped Diamond Multimedia weather the storm. This highly publicized legal dispute is a large part of why the Rio PMP 300 is highly regarded as the first MP3 player in existence.

The Rio name remained active and competitive for seven years. After all the legal disputes, Diamond Multimedia eventually became SONICblue, but was eventually sold to D&M Holdings. The original Rio struggled to maintain sales along with several other SONICblue branded products. Despite a brief comeback with a popular line of Karma and Carbon MP3 players, Rio's patents, applications, and engineering designs were again sold – this time to SigmaTel, a prominent chipset manufacturer. D&M Holdings finally announced that no further attempts would be made to sell the Rio branded player. The trademark of this historic device would no longer change hands, closing the door on the MP3 player that may have been responsible for all the perks we enjoy with an iPod.

Nakamichi Products and Media Players

One of Apple's biggest rivals in the multimedia player industry is company that has been around for a long time — one that was literally established on the forefront of technology. Nakamichi is historically known for creating innovative cassette tape players of the highest quality. Though established in Tokyo, Japan, Nakamichi is now based in Singapore and affiliated with Grande Holdings, a powerful conglomerate based in Hong Kong.

The first of the Nakamichi-branded products was released in 1972 — home audio gear that introduced the world to its first three-headed cassette player. The Nakamichi took a giant leap into the future its SoundSpace system. Over the years, this company has been synonymous in manufacturing quality automotive stereo equipment and home theater systems. Nakamichi introduced its own line of DVD video products in 2006.

The Nakamichi SoundSpace2 Digital Audio Player/ Recorder

Here is one device that exceeds all capability of your average MP3 player. The Nakamichi SoundSpace2 raised the bar with one of its newest inventions. Not only does the unit play music — it records it as well. This extraordinary portable media player fits right in your shirt pocket and is so lightweight that you will forget it is there. The Nakamichi SoundSpace2 has a user friendly interface, multi-line LCD display, and a built-in amp for high-quality output. The package is topped off with a pair of comfortable headphones that provides some of the best audio playback available. This unit is built to last, storing music in solid-state.

Since the memory does not utilize moving parts, the SoundSpace2 is skip-resistant regardless of any strenuous activity you put it through.

Nakamichi's SoundSpace2 digital audio player is highly compatible. This device will detect and decode MP3, WMA, ADPCM, and several other file formats. This is all done by a low-powered RISC processor and firmware that can be constantly updated to anticipate future file formats. All recordings are stored on a very stable, removable, SmartMedia card. Just one 64 megabyte SmartMedia card can store more than an hour of quality music in MP3 format, or more than four hours of recorded media. SoundSpace2 comes with a Windows-based media program that allows you to easily encode and convert unprotected MP3 and WMA files to be compatible.

The Nakamichi SoundSpace2 features a built-in microphone and ADPCM encoder. This enables you to record at anytime and anywhere. Your voice and music recordings can be integrated on a single SmartMedia card and the files will not conflict as they are stored in their own directories.

This Nakamichi portable digital audio player is just one in many great products from the SoundSpace2 line — some of which include the SoundSpace2 Stereo Music System. The device is versatile to say the least, acting as an AM/FM radio, alarm clock, and dictation machine. Voice and music recording functions help it stand out as an amazing product. On top of that, this edition of the SoundSpace2 can compete with feature for feature with an iPod in the way of MP3 capability.

I think it is safe to say that the Nakamichi company may have an advantage over Apple in some areas. As they continue to excel in the portable multimedia industry, players such as the SoundSpace2 may eventually set new market standards.

The Microsoft Zune

The last device to be covered in this chapter is perhaps the iPod's most formidable rival — the Zune. The concept here is perfect. It is another classic case of Microsoft versus Apple. Let us begin to compare the two.

Brief History of the Zune Digital Audio Player

Similar to the iPod, the Microsoft Zune receives the aid of several digital media products and services that enable the device to perform at its best. Some of these components include various software and the Zune Marketplace (Zune.net), Microsoft's version of an online music store.

The Zune primarily comes in three styles and all are capable of playing music, videos, and displaying photos. These devices also have an FM radio tuner. A Zune can share files with the Microsoft Xbox 360 game console through wireless connection via a USB cable. It can also be wirelessly synced by way of a Windows computer.

Microsoft released the first Zune digital media player on November 14, 2006. On October 2, 2007, Microsoft introduced versions 4, 8, and 80 of the Zune. These newer models came with the wireless-sync capability, an updated touch-style navigational pad, podcast support, update sharing licensing, and full support for DRM-free music.

Popular Zune Models

Zune 4 and 8

These models make up the Zune's second generation. Microsoft aptly named these devices according to corresponding storage capacities of 4 and 8 gigabytes. These flash-based units were much smaller than the first Zune, comparable to Apple's iPod Nano.

Zune 30

This was the first Zune model to offer a large storage capacity. As the name implies, this version contains a 30 gigabyte hard drive and also features a 3-inch display and a basic directional pad for easy navigation.

Zune 80

The Zune 80 was considered a major upgrade from the 30 model. The most apparent difference being the amount of storage space.

All Zune models received major improvements by way of upgrades to software and firmware.

Popular Zune Accessories

The Zune digital audio player comes with ear phones, a USB cable, and a stylish carrying case. Add-on accessories include:

- ♫ **Chargers:** AC wall adapters, car audio adapters, external battery adapters

- ♫ **I/O adapters:** FM transmitters, headphones and ear-buds, USB cables

- ♫ **Dock connectors:** for charging, external speaker system, and hands-free capability

- ♫ **Protective Casing:** display covering, hard or cushioned form fitting protection cases

- ♫ **Carrying Cases:** arm bands, belt clips

Zune 80 Gigabyte Model: In-Depth Look

After developing all the great features, Microsoft's biggest challenge was creating a device that could match the iPod in capacity. The Zune 80 gigabyte model would be designated for the test. Consumer reviews on this digital media have been favorable and this was expected for the most part.

Software for this Zune model loads up fast, importing your audio files and photo albums from other applications is simple, and even files from an iTunes library can be transferred.

This version of the Zune is smaller and slightly thinner than previous models. Navigation is a bit different as the pad as been altered to allow you to command the device with a light touch or tap. The 80 gigabyte Zune has also made improvements to the headphones. These fit comfortably into the ear and actually produce a better sound quality than those standard with the iPod.

In the end, this higher capacity Zune is comparable to the iPod in most aspects. Apple and Microsoft have been rivals since then the technology war of PC versus Mac. This rival is sure to continue and become more interesting as the Zune Marketplace rivals the iTunes Store and the Zune portable digital media player battles it out with the iPod.

Zune: Basic Interface

One of the iPod's most beloved features is the interface — a navigational system that can be learned quickly and is simple to use. For all of you wondering just how the Zune player matches up in that department, the answers are in this section. Many MP3 players have standard features that are relative to other brands, and the same holds true with iPod and the Zune. There are quite a few modifications that are significant.

Directional Pad: Learning to effectively scroll with this device is somewhat more of a challenge. The directional pad on a Zune is certainly much different than the iPod's click wheel and touch-sensitive feature. This one comes in the shape of a wheel but is mostly composed of functional buttons. The buttons within the circle are used to navigate. To the left of the pad is a menu button and the Play/Pause button is to the left of the circle. A hold switch is located at the top of the device — basically in the spot you find an iPod's hold switch.

You can navigate through your menus by firmly pressing up and down on the buttons within the circle as opposed to rotating an iPod's click wheel. Overall, the Zune 80 has made drastic improvements to the device's navigational system.

Inside the Zune: After becoming familiar with the directional pad, you will find that the make up of this device is rather unique. The patterns reverse the iPod concept with white text on black menu screens. Highlighted bars fade in and out in white and gray as you access different levels of the menu. The features make for a great visual effect. The Zune even allows you to customize menus with decorative wallpaper. This may be a good idea as the basic screens are rather bland when no action is taking place. There are no dividers to separate the top, from the side, from the bottom; there is just plain text.

Artwork: Here is a feature that really stands out. The Zune displays artwork for your artists in a full-screen view. It also leaves a bit of space at the bottom of the screen to indicate the status of your battery. Overall, the artwork is more advanced than the iPod — largely due to the size of the Zune display screen.

FM Radio Tuner: FM radio capability is another quality asset of the Zune portable media player. Here you get the futuristic fading menus and bright, vivid numbers that display your FM radio stations. Frequencies can be preset for North American, European, and Japanese stations.

Wi-Fi Capability: Here is an area of the Zune with its share ups and downs. The Wi-Fi feature grants you the power to wirelessly loan music to other users. This option is available up to a full day's time. From there, the recipient can purchase content from the music store.

While this feature is unique, I just do not know how useful it is. If you happen to run into another Zune user, it is more than likely that they will already have access to the music store. It can be useful as the Wi-Fi feature allows you to remotely access audio files from a compatible Xbox 360. The drawback here is that a pricey adapter is required to make this happen and the game console itself is rather expensive to begin with.

Notable Mentions: The Zune 80 is larger than the 80 gigabyte iPod Classic. It is also constructed of plastic which may raise the question of physical durability. The Zune gives a user many options for viewing photos. It will automatically display how many photos are in that particular folder, but this can be considered somewhat of a distraction as additional space on the screen is claimed. On the positive side, there is a nice built-in LCD display, great sound effects, easy access to the radio feature, and much more.

iPod versus Zune: The Verdict

Considering all the advancements Microsoft has made in technology, I think it is safe to state that Zune will be around for sometime. So which device is better? That depends on the user. The Zune may be ideal for someone who makes that his or her first MP3 player. However, the dedicated iPod fanatic may find its features and functionality inferior.

Regardless of how quickly the Zune has advanced, I feel this device still has a lot of catching up to do.

While the digital audio players in this chapter will always be compared to the iPod, Apple's signature device will always be the Zune's true rival.

There is always room for improvement, but Microsoft did a good job for the most part.

After years of trailing in Apple's footsteps, the Zune emerged from the shadows and quickly establish a reputable name for itself. It is clear that Microsoft learned a lot from the iPod and nearly every other portable player that came before the Zune. They stayed away from the huge, clustered interface of earlier players and introduced a few exclusive features as well. It should be interesting to see how advanced the Zune will eventually become.

Case Study: Michael Daniel Jr.

Michael Daniel Jr.

Aspiring Artist/Songwriter

I have been writing music for 12 years and recording for seven years. Some of the artists that influenced me the most from the past are, of course, Rakim, Onyx, and Guerilla Funk. And as far as some of the present artists that influence me, I would have to say Lil' Wayne, Lupe' Fiasco, Stretch Money, and just about anyone with something REAL to say.

What led me to become an iPod user was mainly its ability to store a COLLECTION of music and use the MP3 format as opposed to the traditional CD player with the data disc.

The first iPod model I owned was the one gigabyte iPod Nano. To me, the best feature of all iPods is the Touch Sensitive Wheel. I get the most enjoyment out of my iPod when I am writing music. I like to upload my instrumentals onto the iPod and listen to them there so I can hear all of the instruments and samples in the song.

Actually, I have a new computer so right now I only have about 48 to 50 hours of music on my computer, although on my old computer I had about a hundred and some odd days of music in my iTunes library. The artists that get the most plays on my iPod are Lil' Wayne, Jay-Z, and Dipset.

I am not really familiar with any other brand of MP3 player, except for the Zune. I do not think it really matches up to an iPod.

If I had to recommend a model to someone looking to buy their first iPod, it would be either the 30Gb Video iPod or the Video iPod Nano.

Options Other than iTunes

When it comes to online entertainment, there are literally hundreds to thousands of Web sites that exist solely to satisfy your crave. Whether it is music or video content, the Web is sure to have it waiting for you.

The market that makes up online multimedia content is just as fierce as the relative digital audio player industry. While companies work vigorously to keep pace with the iPod, attempts are simultaneously being made to effectively rival the iTunes Store and the iTunes program itself. Some of them match up quite well in the way of capability while others are severely lacking. In this chapter, you will become more familiar with the online stores and multimedia applications that are most commonly used as alternatives to iTunes.

RealNetworks: Media Players and Online Music

RealNetworks Inc. has maintained a solid reputation in the multimedia industry. This company withstood Apple's toughest blow and evolved right along with the fast-paced market. It has been an innovative approach that has kept RealNetworks thriving from the beginning, specifically in the midst of the DRM era.

Harmony Technology, RealNetworks' DRM translation service, was the first of its kind. This was a clever strategy on their part. It was a method of extending their services to other manufacturers in the industry. With Harmony Technology, people who purchased content from the RealPlayer Music store were able to play that music on various MP3 players — this included the iPod.

This was the result of two major disagreements amongst both companies: Apple's smothering ownership over music by way of FairPlay DRM and Apple's unwillingness to make their iPod compatible with the RealNetworks service. With such a dominant hold over the market, Apple felt no need to make partnerships with RealNetworks or any other entity.

At the time, RealNetworks lacked a digital media player of its own. This forced the company to rely on licensing deals with other device manufacturers, with Creative Technology being one of the first. After forming a strong alliance, RealNetworks created a stable online music store that supported more than 70 portable digital audio players including devices from Creative, Palmone, Rio, and all generations of the iPod.

Even without device to call their own, RealNetworks still managed to capitalized on relationships with others in the industry, making their service a great alternative to the iTunes Store. Aside from being highly compatible, RealNetworks also has two popular services that add credibility to its legacy: RealPlayer and Rhapsody.

Review of RealPlayer 10.5

RealNetworks stood amongst Apple and Napster with the introduction of RealPlayer 10.0 — their version of an online media store also offered tracks at 99 cents per download. The 10.5 version came equipped with the highly anticipated Harmony Technology. This player is able to support a variety of file formats including AAC from iTunes and protected WMA that typically came from the Napster service. RealNetworks also introduced a couple of new codes with RealAudio and Real Video, and these components improved the quality of overall playback.

After getting inside RealPlayer, you may find the interface to be somewhat congested. Many of the controls are right in front of you; a factor that may confuse some new users. One plus is that the RealPlayer Music Store has been smoothly integrated with the application. This will show you detailed information on songs, artists, and helpful recommendations.

RealPlayer kicked things off with a promotional effort that offered certain songs for 49 cents. They also have a section, called the Rolling Stone Top 10, that offers reduced prices on tracks.

This category is updated on a weekly basis. Aside from browsing the RealPlayer Music Store, users can easily create playlists and custom mixes, rip and burn CDs, and listen to thousands of their favorite radio stations.

RealPlayer 10.5 was known to endure occasional performance issues — the design was not flawless. However, the integration of Harmony Technology made way for extended capability with a wide range of portable media players. This version from RealNetworks was instantly embraced by the market and considered to be a worthy upgrade.

Rhapsody

Rhapsody is a well known resource for acquiring legal music. This RealNetworks product is an online store with a robust selection of music that quickly increased from 375,000 to more than 3 million songs. Instead of software being downloaded to your computer, audio files are instantly accessed from the Web site. You can easily select and purchase your music and then organize them into custom playlists. You listen to music directly from the Rhapsody Web server — this will inevitably save a ton of space on your hard drive. With the iTunes program, you will notice a substantial amount of shortage in disk space after filling up the library.

Rhapsody requires a member account to access music from the online store. Memberships can be purchased for a monthly fee of $9.95 or a quarterly fee of $24.95. Taking your experience from the Web server can be done by downloading each track for an additional 79 cents. These CDs can then be burned and even converted and synced to portable media players.

Rhapsody offers a generous 14-day free trial offer to get you better acquainted with the system before signing up.

The Downside

There were a few complaints about the ownership of music upon Rhapsody's introduction. In the beginning, there were restrictions concerning the duplication of certain content — a factor that fueled a conception that the majority of the songs on Rhapsody could not be burned. Since then, many of these restrictions have been lifted, making for a more enjoyable experience on the Rhapsody online music store.

Others have complained about the cost of service. This has mostly come from individuals mismanaging their subscriptions. You can simply choose to access one track per month, or get your hands on all Rhapsody has to offer in the highest quality of audio streams. On the other hand, someone who has intentions on downloading his or her purchased content can find this service to be rather expensive and not as worthy considering the additional fees involved.

Rhapsody Improvements

The iTunes Store's dominance only improved RealNetworks' Rhapsody brand. Rhapsody now stands as a well developed product that offers many of the perks you will find in the iTunes Store. Videos have been added along with a catalog that contains a plethora of online radio stations with themes from "50's Rock & Roll" to "Metropolitan Opera Radio."

One of Rhapsody's best new additions is the "Lyrics" feature. This has been embraced by the public as everyone wants to know their favorite song, line for line. Simply type the name of the track into a search box and Rhapsody will display the lyrics. Of course, lyrics for all songs are not currently available. The good thing is that Rhapsody updates the Lyrics feature on a regular basis.

Rhapsody is affiliated with the prominent cable stations MTV, VH1, and CMT. These companies all of have their own sections that feature blogs, news in the music industry, the hottest artists, and even playlists created by the stars.

So how does Rhapsody match up against iTunes? Well, the value of their monthly package all depends on the individual. Someone with a lot of free time on their hands will have a ball with the plethora of music offered at one low price. The downside is the additional fees for downloading and taking complete ownership of the content. Those who sign up for their unlimited service may find this to be of excellent value — this membership is more suitably compared to the iTunes Store. This option allows you to listen to all the music you want and save a few dollars as many of the songs you download may average out to less than 99 cents.

The biggest knock on Rhapsody is its online dependency — should anything happen to your online connection, all of the content you purchased and organized on their server will be inaccessible.

The interface of Rhapsody is vivid and user friendly. While the setup sort of mimics the iTunes Store, RealNetworks has done a good job understanding public demand and making the most of their resources. As long as Rhapsody is around, Apple is sure to have a worthy competitor to keep an eye on.

After years of competing, RealNetworks may have finally ascended to Apple's level of capability and quality. With a solid combination of Rhapsody, RealNetworks and evolving line of MP3 players, this company has cemented their name in the industry.

MusicMatch

MusicMatch is certainly one media player that can be considered the first true rival of the iTunes program. At one point in time, this was the default software used to store and sync music to PC-based iPods. As Apple branded products dominated the market, iTunes took over responsibilities for all iPods, but MusicMatch continued to improve on their media player and eventually developed an online music store of their own.

Review of MusicMatch JukeBox 10.0

MusicMatch JukeBox version 10.0 is the brand's latest attempt to competitively rival both iTunes and the iTunes Store. The latest edition comes with notable improvements to the graphical user interface by offering in-depth customization. MusicMatch Radio, formerly known as MusicMatch MX, made drastic improvements on the program's broadcasting capability. They also took this up a notch by restructuring the Artist Radio function, formerly known as Artists on Demand. Interfaces of both radio functions are now much faster with far less bugs and system errors. Another notable radio upgrade allows you to view frequencies by presets or by customized preference.

Other program upgrades such as the "Auto DJ" feature certainly gives this jukebox a boost in the way of overall capability. This edition exceeds MusicMatch 9.0 by far with a subscription service to MusicMatch On Demand. This feature is easily accessible and capitalizes on a world that now seems to need everything "On Demand." All songs on the MusicMatch Music Store come at industry standard prices of 99 cents per track.

Users of MusicMatch 9.0 will instantly notice the updated look and feel of the newest version. The default color of the display screen is a dark blue and other program templates can be downloaded. Music can be viewed by Album, Album Art, Rating, On Demand selections, and many other customizable views that you will find in the convenient drop-down box. MusicMatch also has a viewing feature that goes beyond ID3 tag criteria. The program allows you to customize a view that displays the most recent WMA tracks, to burn or play, among many other possibilities. Similar to iTunes, music tracks can be easily dragged from the library and dropped into your playlists — even those purchased On Demand. This new and improved interface lets you modify track information right from the view within your library. With MusicMatch 10 Plus you can even change information for multiple songs in a few clicks.

MusicMatch JukeBox 10 truly excels at covering the basics. The On Demand features give users unlimited access to a huge library of more than 40,000 albums and 800,000 songs all for a monthly subscription priced at $7.95 per month. This price is based on calculation of the required annual subscription. This version of MusicMatch comes with streaming capability, a popular new method of accessing media. Users are able to customize their streams with a personal touch with features not found with services like Napster version 2.0. This PC-based program has a much better audio ripper, burner, and music manager — traits of a high quality, well-rounded all-in-one player. MusicMatch's extensive list of new features is enough to distinguish it from competitors, but the best addition may be its On Demand Service.

The Auto DJ feature is similar to the Smart Playlist function you will find with iTunes. This enables you to customize your playlists based on criteria of your choosing. For example, you can drag and drop an artist such as the late, great Jimi Hendrix into the Auto DJ window and set the number of tracks or base the playlist on criteria such as file minutes, hours, or size.

From there you can configure the playlist to determine relative artists, set the source to be either the internal library or On Demand, and assign a level of popularity to the music. The artist used in our example is liable to prompt relative results with artists such as Pink Floyd or Led Zeppelin. After learning the ins and outs of Auto DJ, playlists can be made in a matter of seconds. This is a great way to keep your music fresh and entertaining. The MusicMatch Auto DJ will give you up to 256 megabytes per playlists. That many gigabytes is more than enough music.

Following years of tradition, MusicMatch JukeBox 10 is available in two different versions: the free downloadable basic package that comes with many newer PCs or the feature-rich Plus edition that is priced at $19.99. While often compared to the Apple product, both versions of MusicMatch are compatible with the iTunes Store, but the basic version requires an upgrade to enable this. Only users of MusicMatch JukeBox 10 Plus get the luxury of SuperTagging. This addition analyzes your audio tracks and accordingly updates the ID3 tags based on criteria such as the filename. Plus will also allow you to burn music much faster than the basic version, with duplicating speeds of 48x as opposed to 8x. MusicMatch JukeBox 10 also allows more efficient management of file folders along with a new set of right-click functions that give you more options with your music.

Navigating through On Demand music can prove to be quite enjoyable thanks to the array of well developed screens and subwindows that link to relative sections. Contrary to previous versions, MusicMatch JukeBox 10.0 is designed in a way that makes finding your individual tracks, albums, and artists a painless procedure — both in the program library or the online music store. Relative links are abundant with much more than your typical interface. After experimenting with the program for a while, you will quickly learn how easy it is to add personal flair to your multimedia experience.

Subscribers of the MusicMatch store are able to create and store music in the popular MP3 and WMA formats. The service also provides you with an automatic daily playlist based on relative music you frequently listen to. One of MusicMatch's most unique features is the ability to remotely access On Demand songs and playlists from any PC operating with version 9.0 or higher. You can also e-mail content to friends and they have the freedom to listen to this music up to three times before being required to subscribe to the service. The only requirement is that the recipient must also be running MusicMatch JukeBox 9.0 or higher.

Similar to the iTunes, MusicMatch operates on a well designed, stand alone application as well as the online store. Users can browse and preview music free of charge — after finding the desired content, all that is required is to complete the one-time registration form which does not take much time to complete.

Offering more than 800,000 songs and counting, the MusicMatch catalog is racing on the heels of the iTunes Store. While several files are supported, the basic format for MusicMatch is WMA — the equivalent to iTunes' AAC files. Content can also be purchased by redeeming MusicMatch Gift Cards that range from $10 to $300.

Content purchased from the MusicMatch music store downloads effortlessly into the application on your hard drive. It also syncs to WMA compatible media players with no complications. In addition, recent upgrades of the MusicMatch media player have been designed to support the iPod device.

MusicMatch Summary

The Upside: MusicMatch JukeBox 10.0 made an extraordinary upgrade to its user interface — Auto DJ, On Demand capability, remote access, shared playlists, and advanced CD ripping and burning solutions are a few of many features that stand out. This application also has the ability to securely sync content to a variety of MP3 players and supports a variety of plug and play applications and devices.

The Downside: You have probably noticed that there was no mention of MusicMatch's ability to handle podcasts and video content — that is because it does not. The name says it all — MusicMatch is the ultimate jukebox for all of your music related needs. In this sense, it is comparable to iTunes but suffers in universal value.

Regardless of what the system lacks, MusicMatch is deservingly credited for its part in pushing the multimedia industry forward.

How to Sync Your iPod from MusicMatch

While content purchased from the iTunes Store is exclusively synced from iTunes, there is a way to sync and manage your iPod with the MusicMatch application. Follow these directions:

1.   First, click the "Start" button on your desktop screen.

2.   Navigate through the menu and double-click "Control Panel."

3.   In this menu, find and select "Add or Remove Programs."

4.   Navigate through your list of programs and highlight "iTunes."

5.   Proceed to uninstall the iTunes application.

6. Next, highlight and uninstall the "MusicMatch iPod Plugin" component.

7. Select and uninstall "Apple Software Updates" and other iPod or iTunes related content that you find listed.

8. Now, select and uninstall the "MusicMatch Jukebox" application.

9. Close the "Add Remove Programs" window.

10. Close the "Control Panel" window.

11. Restart your operating system.

12. When the computer has rebooted, double-click the "My Computer" icon on your desktop screen.

13. Double click on your local hard drive.

14. In the list of programs and components, locate and open your "iPod Directory."

15. Highlight all the files inside of the directory and delete them.

16. Next, close the "My Computer" window.

17. Now, you will open the "Recycle Bin" from your desktop screen.

18. Inside the Recycle Bin, you will notice all of the files from your iPod directory. Click the "Empty Recycle Bin" icon.

19. Now, reinstall the Music JukeBox software from a CD or download from Web site.

20. When the installation is complete, reboot your computer.

21. When your computer reboots, connect your iPod.

22. Next, open the MusicMatch JukeBox program.

23. From here, you should notice an icon representing your iPod in the source pane. Click the icon and proceed to drag and drop audio files into the device.

Amazon MP3 Store

Amazon.com has been one of the most prominent Web sites for years. It is the ultimate virtual retail center, popular for selling books, music, movies, electronics, and much more. Amazon recently hit the Web with its DRM-less music store — one that would inevitably be compared to Apple's iTunes Store.

Amazon's MP3 store was met with great anticipation as many wondered if they could actually pull it off. While online DRM-free stores was not a new trend, no company besides Apple had been able to market it effectively; a case that can be argued since the iTunes Store was established with the FairPlay wrapper included. Regardless of those who failed before them, Amazon was intent on revolutionizing the Internet by giving users the freedom to do whatever they wanted with their music, while at the same time supplying a robust section of music from popular and aspiring underground artists.

How the Store Works

Similar to their regular outlet, Amazon's MP3 store has an extensive music catalog. You are sure to find almost any artist, song, or album by typing the information into the search box. Results are displayed on a flash-based page that consists of album details, customer reviews, and ratings all integrated from the primary Amazon database. Those that do not prompt relative results can typically be found from the traditional store front in which they are only available in full albums. The selection is impressive and matches up well with music on the iTunes Store.

Music on Amazon's MP3 store is very reasonably priced. Individual songs go for 89 cents as opposed to 99 cents from the iTunes Store. Prices for albums are flexible, ranging anywhere from below $2 to more than $20. Searching through the store myself, I found that the albums, on average, are typically cheaper as well. For instance, Ghetto D by Master P was $9.99 on Amazon and $12.99 at the iTunes Store. In fact, out of all the albums I searched for, none of them were more expensive than the iTunes Store and at worst, they were equally priced. It is apparent that Amazon is looking to give their customers value prices.

Music downloaded from Amazon is also much more compatible than those from the iTunes Store. iTunes AAC file format does not support many devices while Amazon strictly uses MP3, a file format that has become universal for several digital media players.

Downloading purchased music from Amazon requires the installation of their Download Manager — this only applies for full albums, not individual songs. Installation of this software is fairly easy on a Mac or PC. The process basically consists of a few clicks and the files are instantly downloaded onto your hard drive.

Just like iTunes, the Amazon music store allows you to preview tracks before purchasing, a small but valuable feature. The neat thing is that your account for the regular Amazon outlet instantly carries over to the music store — if you have one. If not, signing up is free and the registration process usually only takes a few minutes to complete.

Purchasing your music is also easy. Simply hit "Buy MP3 Album" beside the item and Amazon's 1-Click system completes the order. I found that speeds of the downloads were somewhat slower than the iTunes Store. It took an estimated 15 minutes to download a 13-track album when the iTunes Store downloaded the same album in less than 10 minutes. After each song is downloaded, the files are organized into the Amazon Download Manager, categorized into folders based on artist and album. If you also have the iTunes software on your computer, the Download Manager will store purchased downloads in your iTunes library.

In the terms of a music store, Amazon may have one up on Apple. They had a few old-time classics that I could not find on the iTunes Store and their prices appear to be some of the most competitive on the market. As far as being a complete multimedia outlet, Amazon's MP3 store still has some work to do. Until they build extensive libraries for videos, podcasts, and radio stations, I do not believe they are in Apple's league. Considering all of the products Amazon has listed on their traditional storefront, I look for these upgrades to be made in the near future.

eMusic

eMusic is one of iTune's oldest rivals and has resurfaced to become a worthy competitor. This online music service has actually been around before the iTunes Store and continues to improve year by year. Debuting in 1999, eMusic is known to be one of the first online outlets to offer MP3 music.

eMusic separates itself from most of the competition by offering their music only in the highly compatible MP3 format. This service is also known for its DRM free downloads, eliminating the issue of files being limited due to expiration dates or restrictions on CD burning.

While the service has made a miraculous comeback, eMusic was once severely criticized for its lack of DRM protection. This, and unbelievably low prices, made eMusic unattractive to the four major record labels. This resulted in the online music store specializing in underground artists and brand new genres that previously did not get much attention. Some of the music categories you will find on eMusic include pop, rock, jazz, electronica, experimental, and hip-hop — all of these are brought to you by independent record labels.

What may have tarnished their name in the beginning is exactly what sparked people's interest in eMusic. No DRM files appealed to music listeners who were limited on what they could do with downloads purchased from the iTunes Store and several other Internet outlets. As it stands now, losing out on contracts with the four major record labels was not such a tragedy, but could actually be considered a strategic business move.

As of December, 2006, eMusic had a huge selection of an estimated 2 million songs available for download — the service had sold more than 100,000,000 tracks. These numbers have increased to nearly 3 million songs in the catalog with more than the 165 million downloads sold, 25,000 of those being attributed to audiobooks, the latest feature of eMusic.

The number of independent record labels that keep this service pumping has surpassed 20,000. eMusic works to the favor of the company, artists, and customers who are looking for new, affordable music.

How eMusic Works

eMusic's biggest advantage over the competition is price. The online store operates on a subscription basis which is where the deals come in. When things average out, you pay about 20 dollars per month for about 75 songs. An album with 10 tracks would end up costing you around three dollars opposed to 10 dollars on the iTunes Store. Signing up for an entire year gets you an even better deal; you will pay an estimated 21 cents per song.

The beauty of eMusic is the freedom a user has with downloads they purchase. You can save them, burn them, and pass them along to friends on an unlimited number of computers. These files can be easily transferred to an iPod or any other digital player that supports the MP3 format.

The audiobooks on eMusic are also free of DRM restrictions. These downloads require you to pay and additional fee of $9.99 per month for one book, or $19.99 per month for two books. The audiobook catalog is impressive, considering it has not been up that long. eMusic has already compiled over one thousand books from authors of various genres. The music store also offers a variety of spoken word downloads along with standup specials from both upcoming and well known comedians such as George Carlin.

Why eMusic May Be For You

While surfing through the eMusic Web site, you will learn that many of the same songs can also be found on the iTunes Store. This means people are overpaying for exactly the same track; mainly because, even though eMusic has been around, many still are not aware of its existence.

eMusic will add much value to your dollar. This is certainly attractive to someone who downloads a lot of music every month. Even without the support of mainstream artists, this online outlet has continued to flourish, gathering more satisfied subscribers.

Another great quality of eMusic are the bonus features that come included with the service. Subscribers are able to re-download tracks at no extra cost and with no explanation. You can even purchase "Booster Packs" that allow you to download additional tracks when exceeding your monthly limit. This allows you to access your favorite songs and audiobooks at any time whether you are at home, on the job, or simply out and about.

The eMusic Web site has built a reputation for being reliable and easy to use. The service is backed by a helpful support team that tends to resolve issues in a timely manner. As eMusic continues to excel, more independent labels are sure to jump aboard and add more popularity to the service.

The Downside of eMusic

While eMusic has now statistically become the second most used online music outlet behind the iTunes Store, there has been some consumer criticism as of late. These issues all stem from their Free Trial offer. The offer states that a user is granted 25 to 50 MP3 downloads with no commitment to pay anything. Customers are now finding that they must first subscribe, be charged for a month's subscription, and then receive their free MP3 downloads.

This is drastic change that many faithful eMusic users just cannot explain. There has been much speculation of how the Free Trial offer no longer exists but is still being advertised by eMusic. So if you do choose this service as your portal for online music, beware of the Free Trial promotion as you may be instantly charged. The good thing is that you are still able to cancel the service at any time.

There has also been some complaint about the service's upgrade system. Sending details concerning how and when to upgrade your account was always natural. It seemed as if the service tried to go advanced by implementing a link in the message that allows a customer to instantly upgrade by clicking it. Being that no invoices or warnings are returned, several customers were unaware that they were being charged for the upgrade. Regardless of what the service offers, this is not an attractive method of eCommerce.

Another problem with this online music outlet concerned the eMusic Remote feature. This was an alternative to the download manager, which had actually been working for a number of years. The eMusic Remote was a bulked down version of a Web browser that allowed users to manage music they purchased. The issue was that many people had a difficult time using it, causing a few loyal customers to cancel their accounts. This is one problem that eventually grew so large that the eMusic support team had a rough time explaining it. Since then the company has put the Download Manager back in place while the eMusic remote is used more frequently to manage audiobooks. They have worked a few kinks out of the system.

The biggest downside of eMusic is the fact there are little to no tracks by popular artists. You may find a recognizable artist listed, but when you preview the song it is actually a cover that someone else remade. Gathering a collection from eMusic can also be time consuming as you certainly want to preview most of the tracks before purchasing them. Remember, there are thousands of underground artists you have probably never even heard of.

With any online music retailer there are bound to be a few drawbacks, which always look worse in comparison to the positive features. Do not totally rule out eMusic over their deficiencies just yet. The service has been around for some time, and their extensive list of faithful subscribers means the company is doing something right.

The bottom line here equals savings and quality music. It is possible that you will discover some of your new favorite artists on the eMusic store. Regardless of the negativity and what the service lacks, it is definitely worth checking out. Visit their Web site at **www.emusic.com**.

Microsoft Zune Software

Overview

Microsoft finally became a legitimate rival to Apple in the portable multimedia industry with the introduction of the Zune digital audio player — this competition just would not be complete without software that could be compared to iTunes.

This application also goes by the name of Zune. It acts as a content management system for the Zune device and library. The first version of the Zune software was an upgrade of Microsoft's Windows Media Player 11. The most recent edition was developed from the ground up. Microsoft stressed compatibility with this software, backing up the system with additional decoders for complex AAC, MPEG-4, and H-264 file formats.

One of Zune's best qualities is the player comes pre-loaded with music — DTS, Virgin Records, and EMI Music's Astralwerk Records are just a few of various record companies that have donated complimentary music.

The Zune player is responsible for syncing audio and video files, images in JPEG format, and podcasts to the Zune device. Files can also be streamed to the Xbox 360 game console.

Similar to iTunes, this program organizes the media in your library and makes way for easy ripping and burning of CDs. Album artwork and information for most songs is automatically filed accordingly into metadata tags, which are similar to ID3 tags in iTunes. Zune also features a messaging system that allows users to communicate and keep track of songs that have been swapped. Shared songs are allowed to be played three times over a three day period before they are required to be purchased from the Zune Marketplace.

The program requires the Windows XP or Vista operating system to work and older operating systems will run the application after modifications.

Microsoft's Zune Marketplace gives online music lovers another great alternative to the iTunes Store. The Zune store is nicely integrated into the program and gives users instant access to millions of songs in the most compatible formats. The Zune Marketplace includes everything you would expect from a multimedia outlet — genre categories, top songs, hot artists, new releases, and featured content. The store has quickly become a popular source for independent artists to gain exposure and sell their music — you will find many more independent artists here than on the iTunes Store.

Music on the Zune Marketplace is set at the typical industry price — 99 cent per track and around $9.99 for complete albums. The store also accepts Microsoft Points, which allow you to purchase content without using a credit card.

Aside from millions of songs, the Zune Marketplace offers users a nice collection of podcasts, music videos, and a growing catalog of movies.

Microsoft also launched Zune Social, an online music community. This social network is a great way for users to interact, create, and share music. Since music on the Zune Marketplace is free of DRM, users are able to import purchased content to a variety of other portable media players.

The Zune Marketplace has come a long way in a short time. Microsoft definitely got off to a good start with this one. The storefront lacks a bit in the way of content in comparison to the iTunes Store; the background design on some of the pages is dull as well. With that said, this service is quickly gaining popularity as the downloads keep rolling in.

Case Study: Triston McIntyre

Triston McIntyre

Freelance Writer

Tech.Blorge — Senior Writer

I am a young writer, having been published for just over a year now. However, as a professional blogger and writer, I have written daily for that entire time span. Writing initially attracted me simply as a venue to voice my beliefs and opinions; I like the flexibility of writing and the challenge of making a public name for myself.

I have always had a passion for technology; even as a kid I was fascinated by electronics of every sort. As I have grown up in the era of the "techie," I have followed most technology from the earliest stages, such as video game consoles, PCs, cell phones, portable media players, and the like.

Tech.Blorge.com really is a unique Web site and group in that we all work closely in developing both our material as well as ourselves and each other as writers. Technically speaking, I am responsible for writing material, evaluating my peers, superiors and less-experienced teammates, and editing **Mac.Blorge.com**, our Apple-focused Web site. Aside from that, we all share the responsibility of contributing ideas for the future of our team and Web site, and working to be on the cutting edge in the incredibly fast-paced online news world.

In addition to writing consistently for Blorge, I've been writing a column entitled "Stuck in the Middle," a controversial political and social commentary, for Frostburg State University's paper *The Bottom Line*. In regards to additional projects at Blorge, I have been doing work reviewing new products such as video games, software, hardware, and more. I switched from Windows-based machines just over a year ago after being a proficient Windows user with an aptitude for building PCs. I chose a Mac for a few reasons: the most important being the simplicity, security, and dependability of Mac's OS X; additionally, the hardware is cutting edge, sleek and innovative. However, it really is all about the software.

I have been using and following iPods since the first model was released. I remember thinking, "Wow, this is the coolest gadget on the market." Not only that, it also plays music quite well. At the time, I recall the original iPod having a much larger storage capacity than other models on the market; the shiny white case was pretty appealing as well.

Case Study: Triston McIntyre

Oddly enough, though, I have used iPods and known just about everything about them since the iPod's birth — my first iPod was the 5th generation Video iPod. I was given a black model for Christmas one year with my name engraved on the back; it lasted a little over 2 years before it died.

The best feature of the standard iPod is obviously the user interface; no matter how pretty a manufacturer's case design is, what really keeps users coming back is the simplicity of the interface. The latest iPod, the Touch, is really an amazing piece of technology, offering almost all the same functionality as the revolutionary iPhone. One of the best features of the Touch is that the memory is flash-based as opposed to a standard hard drive; this allows for faster processing, and flash-based memory cannot wear down like a standard hard drive.

As a writer, my Macbook Pro is almost attached to me like a third hand at all times, so I usually use my MBP for listening to music; however, when I'm walking to class, in the gym, or taking a long car ride (which is frequently), I am listening to my iPod through Bose ear-buds.

I have exactly 13,249 songs in my iTunes right now. I am a fan of discographies, and my external hard drives pay for my love of many artists' entire works. That number fluctuates as I sometimes remove songs I am no longer fond of, or add artists that tickle my fancy. Right now, my most played artists are Ryan Adams, Sufjan Stevens, Mozart, Chopin, The Yeah Yeah Yeahs, Boston, Norah Jones, and Damien Rice.

It is really hard to say what I would like to see in future iPods because I am still blown away by the features in the new Touch and the iPhone. I would like to see the Touch become standardized as the everyday iPod, and for larger amounts of flash-based storage. I think that more functionality would make the iPod almost a tiny mobile computer, running a full version of OS X. Simplicity was always a selling point for the iPod, and too many more features might be a turn-off to many customers. I think one great innovation would be the ability to share pictures and other content with other iPods directly. Obviously there would be many copyright protection concerns, but I think direct connectivity could be a nice feature.

I can always use an iPod when I am away from my own computer; I think most technology writers would find life a little less enjoyable without an iPod. Right now, I am without an iPod, and I must say I am always a little sad to leave the house without an iPod in my pocket. I know for a fact that owners of the iPod Touch love the ability to write material from an iPod and post to the Internet from wherever there is a Wi-Fi connection; I would love that.

Tips & Tricks

We have covered the basic functions of an iPod and also endured some advanced training. By now you should recognize this device is much more than your typical MP3 player. Aside from the many tasks this device was designed to perform, your iPod can used in several other ways as well. In this section, we will discuss a few of the tips and tricks that make your iPod and iTunes program even more extraordinary.

Adding and Changing Album Art

Songs purchased from the iTunes Store come equipped with album art (the CD cover for the album in which that particular track is found on). When tracks are ripped from a CD, the iTunes program will call upon the iTunes Store to acquire album art for those songs.

While the system is on point for the most part, sometimes it does not get it right. The iTunes Store may not have cover art for that particular album or perhaps the store did not match things up correctly and you end up downloading the wrong cover art.

Regardless of what caused the issue, more than likely you will have a few tracks in your iTunes library that have the wrong album art or none at all. Since your iPod is designed to mimic your iTunes library following a sync, this means your device will also lack the appropriate album art.

Fortunately for you, there is a simple solution to this problem. Changing the album art for your songs can be done in a few ways.

Start out by learning just what tracks are lacking art in your iTunes program.

♫   Select the "View" tab at the top of the screen.

♫   Navigate through the menu and select "Show Artwork."

♫   Next, find a track that you know does not have cover art.

At the bottom left of the iTunes screen you will notice a white box that reads, "Drag Album Work Here." From here you will hop online and find a Web site that hosts album art. Simply use your mouse to drag and drop the selected image into the album artwork box. The iTunes program will automatically update the library and add that cover to your track. In the event that you need cover art for multiple songs or a complete album, I have found another method to be quite useful:

♫   Highlight the tracks in your iTunes library.

♫   Click the "File" tab at the top right of the screen or simply right-click on the track.

Both methods will open menus. Navigate through and select "Get Info." When this window opens, select "Artwork." Now you can use the drag and drop technique mentioned above to copy the album art from another Web site, or use the "Add" tab. This will allow you to browse your computer's hard drive for probable matches.

After learning this, you may be curious to know where the missing album art can be found. The best place to search online would be a music store. The traditional Amazon storefront is known for having one of the largest selections of music available. CD Universe is another good source. After locating the album art, simply right click on the Image and copy and paste the file into the iTunes program.

Of course, having artwork for your music is not a necessity, but it is a great way to personalize your library and make your library stand out.

Locating and Deleting Duplicate Songs

After using the iTunes program for a while, a few duplicate tracks are liable to be sprinkled throughout your library. This may be the case on a shared computer when another user unknowingly downloads songs that are already in the library. You will also find that audio files claim a decent amount of space on your hard drive — anywhere from three to five megabytes depending on the length of the track. This may not be an issue if your computer is working with a large hard drive. However, those working with limited space on an iPod Nano or Shuffle realize the importance of managing capacity. This why it pays to be thorough and occasionally browse through your library in search of duplicate tracks. Finding duplicate songs in your library can be done in two ways.

The first way is to select "Music" on the left side of your screen to open the entire music library. From there you can manually scroll through the list of what could easily be hundreds or even thousands of songs. This method can be so strenuous on the eyes, however, that you may end up overlooking a few tracks.

A more stress free way of finding duplicate tracks is to let the iTunes program do all the work for you. Simply click the "View" tab at the top of the screen. In the following menu, select "Show Duplicates." iTunes will then display all of your tracks that share the same file name and list them in alphabetical order. Remember that just because songs share file names does not mean they are duplicates of each other. A perfect example is "Crazy" by Aerosmith and "Crazy" by Gnarls Barkley. I advise you to examine each track and play them in entirety to make sure you delete the correct ones.

Filtering out Playlists

Everyone has one song they must hear. Even if you created a playlist of good jams, you crave to hear that particular one at the time. Or perhaps you would simply like to find out what track has been added to more than one playlist. Figuring this out is easy. Highlight the song, right click it, navigate through the dialog box, and select "Show in Playlist." iTunes will then expand that box and display all playlists that contain the particular song. This is another simple yet useful tip to help you effectively manage your iTunes library.

Affordable iPod Content on Your TV

As we learned before, video content you purchased from the iTunes Store can easily be displayed on your TV — all you need is an iPod AV cable which can typically be found for between $15 and $20. Apple has stated that these are not compatible with iPod models that have color screens, but this is not completely true. You can successfully perform this task by using the basic mini-jack that extends from a three-plug cable. Select "TV" as the output source on your iPod and plug the mini-jack into the device. Lastly, you will plug the connections into your television set in sort of a reverse fashion — video output for an iPod is designated to red as opposed to yellow, the standard for most devices. To simplify things, you will plug the red cable into the yellow input slot on your TV, the yellow cable into the white slot, and the white cable into the red slot. This is a simple trick that can be performed by anyone with an iPod Video.

Parental Control in iTunes

After surfing through the iTunes Store you may find a fair amount of content that bares the "Parent Advisory" warning. This indicates that an item may contain profane language, or references to sex, violence, or drug use. This content may include songs, podcasts, and radio shows available for download on the iTunes Store. Apple has implemented a system to ease the worry of parents and to allow them to set limitations on what can be accessed. You can either restrict all access of explicit content in the iTunes Store or prevent these items from being displayed on the computer screen. Features such as podcasts, radio, and even the iTunes Store can also be hidden from the iTunes library menu so this does not become an issue.

The Parental Control feature was added with iTunes version 5 but has significantly upgrade with iTunes version 7. Parents now have the ability to restrict video content such as movies and TV shows. You can also set specific rating levels based on age in the Parental Control area. To configure these settings, follow the steps below:

Step 1: First, open the iTunes application.

Step 2: From the "Preference" menu, select "Parental Control." In this screen are these options with boxes beside them: "Disable Podcasts," "Disable Radio," "Disable iTunes Store," and "Disable Shared Libraries."

Below this you will find a drop-down menu that allows you to set ratings for the "United States," "Australia," "Canada," "Ireland," "New Zealand," and "The United Kingdom." From there you can make specific restrictions to the iTunes Store based on movie ratings from "G" to "R" and TV Shows rated "TV-Y" to "TV-14."

Step 3: Check the boxes that apply to your preferences.

Step 4: Next, you will click the "Lock" icon to prevent anyone from changing these settings.

Step 5: You will then be prompted to enter an administrative password only you have access to.

Step 6: After clicking "OK," the menu screen will close and your changes should take effect immediately.

To make revisions to these changes, you must access the "Preference" menu from the iTunes program, click the "Parental Control" tab, click the "Lock" icon, and enter your password. This will result in the Parental Controls being unlocked and you can then proceed to make changes.

NOTE: Parental Controls must be configured for each account you want to place restrictions on. Also keep in mind that radio shows found on the iTunes Store are not originated from the Apple company. Many of these broadcasts are strictly intended for adult audiences so parental discretion is advised.

A Smarter Sorting System

Here is a neat little trick that can help you to organize your library like a record store. You may want to have a song categorized by "Jimi Hendrix" or by "The Jimi Hendrix Experience." You can do this for an artist, the title of a song, or a complete album. This is done by tampering with the "Sorting" options that are found within the "Get Info" menu.

The drawback is the Sorting options can only be accessed on a per track basis, meaning this could become a time consuming task. The benefit is you will know exactly where to find your music in the iTunes library or on your iPod.

YouTube on Your iPod

If you have not heard, YouTube is quite the online phenomenon. It is essentially a video-based community where users are able to post all types of personal content, from small clips to full length videos. The following tip will show you an easy way to get any YouTube video you desire on your iPod.

I recently ran across a neat Web site by the name of **www.vixy.net**. Their service gives you a way to place videos from YouTube on a computer, iPod, or one of several other wireless devices. Simply copy and paste the URL for that particular video, specify your file preference, then download the code and place it into the iTunes program. From there you can sync it just as you would any other video content.

Quickly Put Your iPod to Sleep

Here is a handy tip for your iPod Classic or Nano, it is a faster way to put the device to sleep with a simple tap of the finger. Normally you have to press down on the Play/Pause button. This can be avoided by adding a "Sleep" option to the main menu of your iPod. To do this, navigate to "Settings" and select "Main Menu." Toward the bottom of the list you will find "Sleep." Select this option and add it to the main menu. You can now use this method anytime you want to quickly shut down the device. Simply tap the Play/Pause button and the device will instantly turn off. This is an effective way to preserve battery life.

Connect to Wi-Fi with Your iPod Touch

Wi-Fi capability is the latest feature introduced by the iPod Touch. To make this wireless connection to the Internet, you must first be in an area that permits Wi-Fi access. This can also be done by purchasing a Wi-Fi router and configuring it to your high-speed cable modem. After settling into a Wi-Fi zone, follow these steps on your iPod Touch:

Step 1: From the main menu, navigate through and choose "Settings."

Step 2: In this menu, select "Wi-Fi."

Step 3: You should then see a list of available Wi-Fi networks — this is what will establish your online connection.

Step 4: If a WEP or password is required, key in the information and click the "Connect" tab.

After following these steps, you should have access to settings of the Internet provider, allowing you to browse the Web just as on a computer.

NOTE: Proximity plays a huge factor in your Wi-Fi capability. If you are positioned too far away from the router, the iPod Touch might display available networks, yet you may endure complications when surfing the Web or cannot get a connection at all. The best advice is to move as close to the Wi-Fi router as possible. Also, make sure there is not anything impeding the signal.

Rebuild the iPod's Relationship with Your Computer

After mastering your iPod, you will find that nothing can be worse than encountering technical difficulties. When the music stops, it is time to get advanced and locate the source of the problem. One of the biggest complaints amongst iPod users occurs when a computer can no longer detect the device. This is bad for several reasons. The first is you cannot sync content to your iPod. To begin troubleshooting this issue, follow the steps below:

Step 1: First you need to make sure the iPod's hold switch is not on. This will not only prevent the computer from recognizing it, but also prevent the device from turning on.

Step 2: Next, make sure your battery has at least half a full charge. If so, use the method discussed previously to reset the device. Once again, you know it has successfully reset when the "Apple" logo is displayed on the screen.

Step 3: If the computer still will not recognize your iPod, allow the battery to drain completely and then retry it.

Step 4: When all else fails, you may need to update the firmware for your iPod. This can be obtained from the Apple Web site.

If none of these tips solve your issues, the problem may be related to factors, devices, or components other than your iPod or computer. The issue between your iPod and computer may be the result of USB or FireWire connection. This is often caused by a problem with the cable connector. The easiest thing to try is to use a brand new or a different cable.

Next, you can unplug all of your USB-based devices such as the printer, mouse, or keyboard in case one of the drivers associated with the device is conflicting the unit. If this is the case, you should remove and then re-install that device. The last thing you can try is to connect the iPod to another computer and see what happens.

How to Control iTunes' Behavior

Here is a tip that gives you a little more control over iTunes. You will notice after plugging your iPod to the computer, your iTunes program will automatically open. For those of you are annoyed by this, here are steps for a solution:

Step 1: Connect your iPod to the computer and allow the iTunes program to open.

Step 2: When the computer recognizes the device in iTunes, click the "iPod" icon.

Step 3: In the iPod menu, uncheck the option that reads, "Open iTunes when iPod is connected."

The next time you connect your iPod to the computer, the iTunes program will not start up unless you open it yourself. Again, this tip is very elementary, but it took a while for me to figure it out.

Using your iPod for PowerPoint Presentations

Here is a trick for the advanced iPod user. PowerPoint presentations have become a common method of advertising products and business plans — an innovative way to get your point across. Instead of lugging your laptop or desktop computer all over the state for business purposes, you can lighten the load by giving breathtaking presentations using your iPod.

This trick is backed up by an iPod's capability to display JPEG files — this includes the cool slideshows we mentioned previously in the book. Your goal is to convert the slides of the presentation into a series of JPEG images. This means each slide needs to become an individual JPEG file. This can be done directly from your PowerPoint application. You can also do this by using software that captures screen images — most video editing programs have this capability. In this case, you would capture each individual slide and save it in JPEG format. From there, you will store these files in a single photo folder and then sync it to your iPod.

To get the best results, it is best to not display the presentation on your iPod — a video projector that shows large images is a better idea. This will require an iPod AV cable or a similar adapter. To display the presentation, select "Photos" from the main menu of your iPod. Next, select "Photo Library" and choose the folder which contains that presentation. Pressing play will initiate the presentation. Use your iPod's click wheel to advance the slides. Your slides will need to be enabled to advance manually to give you control of the presentation. Turn off the shuffle mode to make sure the slides of your presentation are played in sequence. The results are astounding — a vivid, professional-looking PowerPoint all made possible by your little iPod.

Working Your Way Around Copyright Restrictions

Many people may not understand this, but purchasing music from the iTunes Store does not mean you have full ownership of that content. It is more similar to a restricted lease. Earlier in the book, we discussed how Apple has set regulations on what can be done with the music you download. The rule is that the songs you purchased can be played on no more than five computers. This music is also only intended to be played on an iPod. This is made possible by DRM technology. DRM is set in place to protect the artist's copyright over the music. This in turn places limitations on your portable music.

So how does one get around these DRM restrictions? Remember, the iTunes Store has an entire section of DRM free music — iTunes Plus. You can purchase your content from there, though the selection is not as extensive as the regular store front. The music you will find on iTunes Plus is a bit more expensive, but what you are actually purchasing is freedom. You can play this DRM free music on an unlimited amount of computers and any other digital player that is capable of playing AAC files.

This next method is considered to be prohibited, but it works. The trick relies on the ripping and burning capability of your iTunes program. What you want to do here is burn the music to a blank CD, then rip the tracks back to iTunes in MP3 format. When the CD is burned, the AAC files are converted to an audio format used with compact discs — this will remove the protected DRM wrapper. After ripping those tracks from the burned CD, you will have no DRM restrictions to worry about and can play the music where and however you like.

Creating Multiple iTunes Libraries

Individuals who were instantly intrigued by the iPod may experience a bit of dilemma — first you bought an iPod Nano, upgraded to an iPod Video, and just could not resist the new iPod Touch. Your collection of super devices may be impressive, but can also place you in a jam. What will you do if all of these iPods are synced to a single computer? How do you avoid having the same line up on each iPod?

Apple recognized this probable issue and made adjustments with the introduction of iTunes version 7 — an upgrade to the software that allows you to create multiple libraries on a single computer. This lets you assign those different libraries to each of your iPod models. Now you can have an iTunes library dedicated to the music you play on the road, and a completely different one to watch your favorite movies and music videos while at home.

While you can create multiple libraries, the procedure can be considered a secret as you will not find this tip in the help section of your iTunes program. You will begin by closing iTunes. Next, hold down the "Shift" key on your computer (or the "Option" key on a Mac). Double click on the iTunes icon as if you are opening the program. Instead of the program opening up in normal fashion, a dialog box will display. You will notice an option that reads "Create New Library." Click the button and enter a name for the new library when prompted. iTunes will then open, displaying no content. This represents the new library you just created. You can now add your content and assign an iPod to that library. Use this method to open iTunes whenever you want to manage your multiple iTunes libraries.

Making Your Own Scratch-Resistant Screen

One of the biggest complaints about iPods has nothing to do with its features — it relates to scratches. By mishandling your iPod, you will learn the hard way that the screen can be scratched easily. Before beginning, I must state this trick is not recommended but it works. Those of you of who are familiar with the Sony PSP game console know the faceplate is durable, almost impervious to scratches, certainly more than an iPod screen. You should also known that the PSP faceplate is very similar to an iPod in its size. By disassembling both devices, you can switch out the faceplates and create a more scratch-resistant iPod. This would require you to file down the sides of the PSP faceplate to make for a perfect it.

As you can see, this trick takes the skills of an expert. If you do not have any knowledge on how to disassemble the device, this trick may completely wreck your iPod. For those of you that are courageous enough to try, we have listed directions for disassembling your iPod below.

NOTE: Disassembling your iPod requires both a small Phillips and flathead screwdriver.

First, place the iPod face up on a flat service and give yourself a little space to work with.

By turning the iPod to its side, you will see that the color faceplate of the iPod is connected to a metal case. Take the flathead screw driver and wedge it in between the groove. You want to pry the casing upward from top to bottom.

After one side is free, gently wiggle the plate to free the other side.

You will then flip the iPod face down and lift the backplate up — try no higher than a half inch.

Inside the iPod, you will find a circuit board with several wires and components. In the bottom left corner of the device you will find a battery connector with a brown catch on each side. Use your flathead screwdriver to raise both sides of the connector and carefully remove the power strip.

The hard drive of your iPod is blue and connected by a series of wires. Raise this from the case and flip it over the bottom of the device.

Use the tip of your flathead screwdriver and release the catch on the battery connector. This will cause it to lift from the bottom.

Carefully slide the cable from its connector once the catch has been released.

NOTE: Do not disconnect the iPod's hard drive. Even if it detached successfully, these wires are very sensitive and susceptible to breaking.

Switch on the catch from the connector on the side of the iPod — this sends out signals for audio and video external devices such as speakers on a television set.

To gain access to the front of the device, you must remove a set of six screws on the side — this is what keeps the faceplate connected to the metal case. After the screws have been removed, you will then be able to slide off the faceplate from the iPod.

The most trying process of this entire ordeal is toying with the PSP faceplate. You need to cut a hole in the faceplate (making sure it is a perfect circle) to give you access to the iPod's click wheel. From there, you must file the sides of the PSP faceplate to make sure it fits over the iPod's metal case. The safest way to make both sides stick together is by covering the groove with a piece of electrical tape or a long sticker.

Even an expert may find this to be a challenge. Perfectly filing the PSP faceplate can be a time consuming effort. In the end, the modification of your iPod will be very unique. You will be able to see through the transparent PSP faceplate to the circuit board. Most of all, your iPod will be more durable and protected from easy scratching.

Stream Your Music From Anywhere

After creating the ultimate library, you will find it hard to be without your songs. Do not worry, there is way to access your music even without having your computer or iPod with you. There are several third-party Web sites on the Internet that will allow you to stream your entire library from a different location.

One such site is **www.simplifymedia.com**. From this site, you can authorize up to 30 different computers to remotely access your iTunes library. This requires your personal computer to be connected to the Internet.

There are more services that allow you to remotely stream your music. With **www.nutsie.com**, you can upload your entire iTunes library to their Web server. From there, you can access the music on any computer or cell phone. Both services require users to sign up for an account.

NOTE: These services have limitations concerning remote access. For instance, you will not be able to burn content, or sync it to another digital player — this includes an iPod.

Customize Your iPod

One of the best things about iPods is they can be easily customized — this can be done in appearance or from within. There is a cool Web site perfect for giving your iPod a personal touch. On **www.ipodwizard. net**, you will find a variety of community forums that offer great tips for transforming your device. You can also download the iPod Wizard, a small application that allows to get inside your iPod and edit texts on the screens and menus.

For instance, you can change the "Now Playing" text to "Now Jamming." The site also has a number of features that allow you to modify the graphical interface of your iPod. All themes are created by users of the community. If you do not care for those being offered, simply create your own.

Beef up Your iPod Mini

Currently, the iPod Mini is one of the most popular models. The reason is great portability and price. After saving money, you may find that your iPod does not have large enough capacity to hold all of your music. This is because an iPod Mini does not have a ton of storage space to begin with because both AAC and MP3 formatted files can claim a lot of excess space on occasion. It is a common ailment in an iPod Mini, and other models as well. To effectively resolve this issue, follow the directions below.

First, you need to update your iPod or iPod Mini with the most recent firmware. Firmware and software for your iPod can be download for free via the Apple Web site.

This trick also requires the aid of the iTunes program, preferably the most recent version.

You will then use Windows Media Player to rip your tracks, instructing the program to convert the files into MP3 format.

From here, you will go to your music directory and locate that converted music file. Right click the file and select "Properties." You want to make sure that iTunes is assigned to handling the file as opposed to Windows Media Player. If iTunes is set to play the file, click "Open With" and choose the program for the list of available options.

After all this has been set in place, open each MP3 file individually in iTunes and the program will automatically store them into the library. You can then organize the files to your liking by renaming the song, artist, and album. Once this is done, close the program and plug your iPod into the computer. The iTunes software will then reopen on its own and automatically update the information on the device. When files are converted on another program, iTunes compresses the MP3 format and makes way for more space by omitting the original AAC files. I am sure Apple did not intend this, but it is a rather useful tip to get the most out of your iPod.

NOTE: Mac users can try this by converting files with the QuickTime application.

Tips for Battery Blues

Though the battery of your iPod is capable of providing hours of audio and video playback, it will weaken over time. There have been several complaints about the difficulty of replacing the battery. Even though battery life intends to improve with each new iPod model, there is a chance of running across a less than stellar battery. On top of that, no battery is eternal.

Of course, Apple has you covered when your battery begins to die out. You can usually find a brand new battery for around $50 to $60 on their Web site. While you can have confidence that the battery will be of quality, there is also a more cost effective way to do this.

NOTE: Changing your iPod's battery is not as challenging as it may seem. At the same time, beginners should take heed. If you are intimidated by installing and removing components such as RAM memory in a computer, you may want to avoid this. Frustration can easily lead to scratching up your iPod or worse. Opening the unit may also void your warranty with Apple so you should proceed with caution.

There are numerous Web sites that sell iPod batteries: **www.ipodbattery. com** and **www.pdasmart.com** are two of them. You can find batteries reasonably priced at around $30 to $50, depending on the iPod model. These services offer kits that come fully equipped with all of the tools required to change your iPod's battery. The kit from PDASmart comes with a flexible plastic tool that prevents your case from being damaged. This is a much more secure method opposed to the flathead screw driver used in our previous trick.

After purchasing a new battery, it is time to play doctor with your iPod. Make sure your iPod is off with the hold button switched "On" before opening it. Proceed by inserting the tool in between the groove connecting the faceplate to the metal case. Slide the tool down the side and pry upward to loosen the iPod. While sliding and prying, you unlatch the clips holding the device together and soon, the metal case will slip off with ease.

With the iPod exposed, you will see all its organs — a series of wires and other electrical components. You will also find the battery resting on top of the hard drive. For this next part, you can either use the tool from your kit or a small flathead screwdriver — this may work best here. Using light force, you want to lift the battery from the hard drive, releasing it from the two clips holding it down. After successfully removing the battery, detach its connector from the circuit board.

For the replacement, you should begin by connecting the cables of the new battery to the circuit board. Then, gently lay the battery flat in its position. Make sure your wires are neatly placed. You do not want anything to prevent you from closing the device. The last step is to power on your iPod to see how well it works. If everything seems fine, give the battery a complete charge and jam to your iPod like it is brand new.

How to Unfreeze Your iPod

As an avid PC user, many of you may be familiar with the ghastly blue screen of death — this is when your operating system locks up without warning and leaves you frustrated. Since your iPod runs on a hard drive, it is susceptible to the same behavior. Perhaps the tracks on your iPod are skipping or go silent, or maybe the device freezes up on you. When this happens, there is a great chance that it is something minor that can be fixed in a matter of minutes. To begin troubleshooting, follow these steps:

Step 1: First, you need to assess the problem as best you can. This will help to determine what is wrong and how to correct it. If the screen on your device constantly freezes or will not turn off, you should connect it to your computer or a power outlet. If this corrects your problem, your battery just may be dead and you should refer to the previous tip to address that issue.

Step 2: If the problem persists, there may be other issues. We will keep things simple for now by trying to reboot the device. This will introduce you to an old trick that every iPod user should know: simultaneously hold down the "Menu" and select button. This will cause the iPod to quickly shut down and automatically reboot. The process may be somewhat slower depending on the issue and the capacity of your device. When the Apple logo appears on the screen, you will know the iPod has been rebooted. Now, give your iPod a spin and see how things are working. In many cases, this will resolve your freezing dilemma.

NOTE: If neither of these methods work, you may have to let your battery slowly drain. When it is completely out of power, recharge the device and give it another shot.

Step 3: So we have come this far and your iPod is freezing— do not lose all hope yet. This problem possibly can be solved by reverting to the device's factory settings. Your iPod may or may not have come with the iPod Updater utility. This will automatically search for firmware updates that are available and it will perform a factory reset. If you do not have this utility, the iPod Updater can be downloaded from the Apple Web site. Once you have it, connect your iPod to the computer and select "Restore." This will essentially transform your iPod into a new device. All of your content will be deleted, but this is an effective way of updating your firmware — a common reason of freezing.

If you find your iPod is still freezing after trying all these methods it could be one of three things: a defective battery, a complete failure of the hard drive, or the device may be faulty. iPods are amazing but they are not perfect.

Convert DVDs to Your iPod

I am sure you have saved a lot of space in your house by switching out those old VHS tapes for DVDs. Though the iTunes Store has a huge catalog of movies, you may be wondering how you can save a few bucks by using the DVDs you already own. This may seem impossible, but more and more users have learned how to successfully sync their DVD movies, TV shows, and music videos to an iPod. If you are familiar with ripping DVD movies, this process should be a breeze as they are quite similar. What you need is an effective tool to get the ball rolling — a solid program that will convert your DVDs to the iPod.

QuickTime backs up the iTunes software quite well, allowing you to convert DVDs, but there are some restrictions. For instance, QuickTime is only capable of converting videos that are in the compatible formats of MPEG and MOV.

The best approach here is to use a quality program such as iPod Converter by Videora. You can use this service on a free trial basis to see how it works and purchase it for $30. iPod Converter is convenient as it only claims about six megabytes of space on your hard drive. The utility easily transcodes your original or copied DVDs into MPEG-4 files. It even supports several other popular formats such as MPEG, AVI, and WMV. The program is also capable of converting files TiVo files which is great for recording your favorite television shows.

After using the iPod Converter, you will learn how it easy it is to take your DVDs with you on your iPod. Follow these directions:

1. Download the software from the Videora Web site.

2. Install and run the program.

3. When the program opens, click the "Setup" tab.

4. Next, select your video output source. You may have already saved DVDs on your hard drive in the "My Videos" folder.

5. Click the "Convert" tab.

6. Click the "One-Click Transcode" button.

7. Navigate to the folder or source that contains your video content. Press down the "Control" key as you click to select multiple files.

8. Click "Open" to access the files and the iPod Converter gets going. Patience is the key here. The transcoding process can take as long as the length of the movie.

9. After the video files have been converted, open your iTunes application.

All you have to do from here is to import the converted files into your library, and drag and drop them into your iPod as if you were manually managing it.

The most difficult challenge you will face with this program is waiting during the transcoding process, as you are probably more than anxious to get those movies on your iPod. One of the only files iPod Converter does not support is DVR-MS, a format used by Media Center computers to play recorded TV shows. This program still gives you several options. You can save a few dollars by converting many of the shows that are for sale on the iTunes Store. This requires the aid of another program — MyTVToGo created by Proxure. Proxure was designed solely for converting TV recordings and placing them on an iPod. These two programs will expand the capability of iPod and ultimately make for a more enjoyable experience.

EQ Booster for Your iPod

Overall, the quality of sound on an iPod is very clear and balanced. After plugging in the ear-buds and playing a song, you may have the same impression. On the other hand, an iPod will not add enhancing elements to recordings, leaving certain music flat or distorted. Several rival devices have placed emphasis on this area, implementing features such as automatic bass boosting response.

For the most part, the bass response on an iPod works well. In case you feel the need to enhance it, adjustments can be made to your bass and other areas of music via the iPod's built-in equalizer presets. As mentioned before, the equalizer on your iPod is much less advanced than the one on your iTunes program and it is even more inferior to a few comparable digital players. Since the adjustments are preset and sliders cannot be controlled individually, there is only so much that can be done to a track's output. Getting quality results out of your iPod's EQ may lead to a level of frustration that makes you give up and play your music as it is, distortion and all.

When your music begins to lack quality, there are essentially two routes that can be taken. The first one is the decision to invest in a set of high-quality, iPod compatible headphones. Combing a few technical forums on the Web, I found that the Sennheiser HD 25-1 and the Sony MDR-G74SL are two of the best products to use.

These enhanced headphones are sure to boost your bass, but can also overdue it on a few booming hip-hop tracks. Besides, headphones can be an expensive accessory. They are also limited in functionality as that high powered output cannot be turned off. This puts you in a similar situation and adds a new level of frustration. It would be much more convenient if the EQ on your iPod simply worked the way you wanted it to. There is some good news — by using a third-party utility, issues with bass and distortion on your iPod can quickly become a thing of the past.

MP3Gain

Eliminating the deficiencies in your iPod's EQ is made easy with MP3Gain. This is an open-source application, meaning capable developers have access to the source code and can make various improvements. On top of that, MP3Gain is free to download and use.

MP3Gain is the perfect solution for adjusting the bass and other sound quality issues with music. The program will modify the value to a lower level. Do not worry, MP3Gain will not alter anything else within your file and the quality of your tracks will not be affected. All adjustments of the music can be reversed at anytime.

Why Use MP3Gain?

The volume tag of an MP3 file is set at an estimated 95 to 105 decibels by default. The original settings of your iPod will play without distortion at this level. However, if you increase the bass on the iPod's EQ, the volume of much lower frequencies are boosted beyond this level, ultimately provoking crackling and distortion. The iPod is attempting to output a greater value than the preset limit. In this scenario, MP3Gain will reduce the volume of your music to an estimated 89 decibels, which leaves you a bit of space to make adjustments to other frequencies without exceeding the limit of your iPod's amplifier.

How to Use MP3Gain

By visiting the MP3Gain Web site, you can grab a copy of the program by clicking "Download" at the top of the screen. From the download page, choose the most recent edition of MP3Gain. The download and installation of this software should take no longer than a couple of minutes. To begin adjusting your MP3 files, click the "Add File" or "Add Folder" tab to instruct the program where to pull your tracks from.

You want to make sure that the "Target" value of volume is set at 89 decibels, then click "Track Gain." MP3Gain will automatically analyze each MP3 track that you select, applying the new volume settings. This process could take up to several hours depending on the number of tracks. There is no need to sit around and monitor the process. This may be somewhat of a hassle, but in the end you will find it is worth it.

After the conversion is complete, sync the tracks to your iPod as you normally would. Access the equalizer on the device and open the "Bass Booster" preset. After listening to the reformed tracks, you will find the new decibel level eliminates issues of distortion. You can also go through each individual preset and find the one that suits your tracks the best.

The good thing about MP3Gain is each adjusted song will be equalized. This configuration applies to volume differences between tracks and albums, so you will not have to adjust the volume as much. MP3Gain will determine how loud a user perceives a track to be, as opposed to how loud it actually is. From there, it adjusts the volume accordingly to align it with the rest of your tracks.

MP3Gain is reserved for songs that are in MP3 format. However, there is good news for AAC users as well. The developers of MP3Gain now have information on their home page regarding their latest software, AACGain. As far as I know, this program is still being tweaked. As AAC is the default format for files purchased from the iTunes Store, this program could prove beneficial to several iPod users.

How to Export and Print Your Playlists

There are many ways to organize your music right from within iTunes. The iPod enthusiast may wish to take it one step further by having a documented list of every song in the library. Formulating this content into spreadsheet would be the best way to organize this, but how would you do it?

While I first believed this would require a third-party utility, I learned that iTunes was naturally designed to handle this function. It took me a minute to figure things out, but after some time I discovered an easy way to export songs and playlists from the library as a Text or XML file. You can learn with the following steps:

Step 1: First, open your iTunes program.

Step 2: Next, click the "Music" tab on the left side of the screen.

Step 3: When "Music" has been highlighted, right click it.

Step 4: When the dialog box pops up, select "Export."

Step 5: This will bring up another dialog box — a "Save As" window giving you the choice to save the library as a Text or XML file. I suggest saving it as a Text file as it more universal.

Step 6: After exporting the files, you can then drag and drop it into any spreadsheet application, such as Excel. From there you can organize the date however you choose.

This same technique can be applied to the movies, music videos, podcasts, and playlists in your iTunes library. Simply highlight the appropriate category and follow the directions above.

There is a drawback that you may find frustrating: iTunes will typically export more file-related info than you would like. This could be the track number, date modified, bit rate, and so forth. To keep your list looking neat and orderly, you may want to print it from the iTunes program. Highlight the appropriate category on the left side of the screen.

Next, open the "File" menu and select "Print." The options displayed will be: "CD Jewel case insert," "Song listing," and "Album listing."

Below you will find drop-box menu of themes:

♫   Text only

♫   Mosaic

♫   White mosaic

♫   Single cover

♫   Text only (black and white)

♫   Mosaic (black and white)

♫   Single side (black and white)

♫   Large playlists (black and white)

You can play around with these options to see which one gives the best results. After highlighting a theme, iTunes will display a sample of what it may look like. Personally, I find the album listing option to be the most visually impressive — especially if you have a nice collection of artwork for your music.

eBooks on Your iPod

Previously, we discussed the wonder of audiobooks and how they can be synced to an iPod. Over the past few years, eBooks have become a popular way to enjoy the reading experience. An eBook is the manuscript of a book saved in a PDF file format. This a great, inexpensive way to enjoy books right from your Mac or PC.

For all of you who have to finish reading an eBook right away, there is a handy utility available that allows you to place PDF and other forms of text files directly on your iPod. The program is called eBook to Images, a very useful tool that extracts text from PDF files, and converts them into numerous readable text images. eBook to Images allows you to change text orientation, colors, and fonts. This utility also lets you determine the screen size of your portable media player, then produces images to fit those dimensions. The program is compatible with iPods, cell phones, and most other brands of MP3 players. You can learn more about eBooks to Images at **www.merlinsoftware.com**.

Enhancing Visual Effects on iTunes

iTunes is an amazing program that goes beyond the typical digital audio management system. After a while, you may feel as if the theme is bland and there is nothing going on but the list of your music. Here is a trick to help you spice up your iTunes screen, similar to the colors you see on Windows Media Player. To start the visual show, follow these directions:

At the top left of your iTunes screen, click "View."

Navigate through this menu and select "Show Visualizer." The background of your iTunes screen will become black; the "Apple" will then appear in the middle of it. The show will begin as colorful lines and images dance about your screen. If you grow tired of the same background, here is how you can control the visualizer:

Press C — Shows current effect in various combinations

Press Q and W — Makes adjustments to your foreground patterns

Press A and S — Makes adjustments to your background patterns

Press Z and X —— Makes adjustments to your colors

Press Shift and 0-9 — Saves your current combination of effects

Press R — Displays a random effect

Press M — Turns on user configuration mode and cycles through combinations of effects you have saved, turns on freeze of current configuration, continues to display current effect (also turns on Random slideshow and tracks back to random visual effects)

Working with your iTunes Visualizer will takes some time to master, but you will be able to create breathtaking effects in a matter of minutes.

More iTunes Tricks with Your Keyboard

Not only can your keyboard be used to control the iTunes visualizer, it can also command other aspects such as playing songs and moving through albums. Following is a list of keyboard commands and their resulting actions.

Enter or Return — immediately plays highlighted track

Right + Left Arrow — plays next or previous track

Control + Left Bracket — move to previous page in the iTunes Store

Control + N — create new playlist

Control + O — add a file to the library

Control + Shift + O — import a track, playlist, or file from library

Control + I — opens "Get Info" menu for highlighted track

Control + R — displays where song is located on your hard drive

Control + L — backtracks, highlights to track playing in a playlist

Control + T — turns visualizer on or off

Control + F — turns visualizer into a screen saver when visualizer is on

Control + Z — undo the previous action

Control + X — cut the highlighted track

Control + C — copy the highlighted track

Control + V — paste the highlighted track

Control + B — show or hide artist and album info, online connection will display every album that song is a part of

Control + A — highlight all songs in a playlist or library

Control + Shift + A — deselect all songs in a playlist or library

Control + G — show or hide album art

Control + J — open "View Options" menu for highlighted track

Spacebar — start or stop playing highlighted track

Control + Up Arrow — increase volume

Control + Down Arrow — decrease volume

Control + E — eject CD

Conclusion

After learning about the amazing iPod, one cannot help but wonder how this device will evolve from here. Advancements in technology are inevitable — the iPod is solid proof. Just think back 10 to 15 years ago, no one would have ever believed that it would be possible to walk around with your entire music collection in your pocket. Of course, that was life before the iPod, an invention that proved anything is possible.

So where does Apple go from here? An iPod with enough capacity to hold every single item from the iTunes Store? An iPod that plays pay-per-view events? Here are just a few of the predictions floating up the future the iPod and iTunes:

iTunes Movie Rentals

There is one rumor that has been circulating for some time now. Public speculation came about when a user noticed a heading about a future movie rental system in the customer service area of the iTunes Store. At the end of 2007, *Financial Times* reported that Apple and Fox had reached an agreement regarding the issue. It is also rumored that Apple has similar negotiations on the table with other prominent production companies. As it looks now, movie rentals at the iTunes Store seem to be probable.

The End of DRM

One of the biggest predictions concerning Apple's future in the portable multimedia industry is that DRM will no longer be attached to music. Since May of 2007, the company has been selling DRM-free tracks via the iTunes Store. Both Amazon and eMusic offer songs without the restrictions of digital rights management and have been quite successful. Three of the four major record labels have already signed on with Amazon's MP3 store with the fourth one expecting to come aboard soon.

It is expected that most, if it not all, available songs on the iTunes Store will be free of DRM. This would more than likely eliminate Apple's iTunes Plus section of the store. Considering how fierce competition is and past disputes revolving around the use of DRM, I look for this prediction to go from speculation to fact.

Touch-Free Interface

The Apple company recently filed a patent that gives some idea of where they intend to go with new iPod generations. It details an iPod interface that can be controlled without physically touching the screen. I have a hard time picturing it, but it could be a breakthrough.

While there is really no reason to touch an iPod's screen, you will find yourself doing it rather frequently. Anything that reduces the number of scratches and smudges caused from daily use is certainly a plus.

Keep in mind that Apple files several patents, some of which will never become anything more. It is good to know a little more about the probable road the iPod will take into the future.

Possible Features for Future iPods

The iPod Touch has claimed fame from its unique multi-touch screen interface. It has also been knocked for its lack of capacity. The iPod Touch comes available in eight or 16 gigabyte, flash-based models. It is likely the future will see this model upgraded to a super-slim feel of the iPod Nano and robust 160 gigabyte capacity like the iPod Classic.

Though nothing is for certain, iPod evolution can be analyzed from patterns taken with each generation. Apple tends to take features from newer models and implement them into the lower end iPods. For example, the newest iPod Nano now comes with most of the features found in the iPod Classic. Considering the rave behind the new multi-touch screen interface, it would not surprise me if this became the way of control instead of the click wheel.

There has also been speculation about the iPod Touch becoming Apple's new flagship model — a move that would possibly stop production on all models of the iPod Classic. When you think about it, 160 gigabytes would be much more handy on the iPod Touch as opposed to the iPod Classic. The screen alone would make for a more enjoyable experience when it comes to video content.

What about the iPod Shuffle? This device cannot possibly get any smaller. However, it does have room for improvement in the capacity department. A small screen and a full click wheel would be a plus as well. Creative Technology's Zen Stone Plus and RealNetworks' Sansa Clip both come with color screens, have larger capacities, and are cheaper than the iPod Shuffle. Look for Apple to make upgrades to this model in the near future.

The Wrap-up

Regardless of what the device may lack, or the number rivals competing for its throne, the iPod still remains the most preferred MP3 player. Apple's ingenious creation is one that has crossed demographic barriers, changing the face of the portable multimedia industry on a global scale. People of all ages and occupations have found this device useful, from teachers and business moguls to athletes and celebrities.

Although the iPod has become extremely popular, there is still a fair amount of people who have yet to experience one. There is also a number of iPod users who have yet to reap the maximum benefits of the device — the ability to store and play a tons of music is good enough for them. This is in large part due to the complexity of technology. Let us face it — many of us stared at our DVD player as if it were a high tech space gadget when it first hit the market. Anything we are not familiar with can prove to be intimidating in the beginning.

While small in size, the iPod is a rather sophisticated device. Without practice and a bit of training, learning the ins and outs can be a challenge. The good thing is the iPod was designed with the user in mind, building a solid reputation on simplicity and its array of amazing features. After becoming familiar with the interface and how the iTunes program works, you will find that mastering an iPod is as easy as learning to manipulate your DVD player.

Since this book covered numerous topics, let us now give a brief recap of what was discussed:

- ♫ How the iPod came about

- ♫ The basic controls and features of your iPod

♫ All the available iPod models

♫ The wonder of iTunes and the iTunes Store

♫ Advanced work with your iPod and iTunes software

♫ Many available accessories to enhance your iPod experience

♫ iPod's most competitive rivals

♫ Alternatives to iTunes and the iTunes Store

♫ Tips and tricks to get the most out of your iPod and iTunes program

Apple has made profitable history with the iPod, one that has left a lasting impression on the industry. Though this book is not guaranteed to make you an expert, it will help you learn the basics while teaching a few advanced techniques as well. More importantly, this book explained in detail how Apple offset the portable multimedia revolution, a factor that has proved beneficial to both users of the iPod and rival MP3 players alike.

I conclude this book by providing you with a list of viable resources where you can find more information, or even purchase an iPod or some iPod accessories.

http://www.apple.comwww.apple.com

You will find the Apple Web site to be the most reliable source for iPod and iTunes. This site contains everything you need to know about available models such as features and technical specifications. The Apple Web site is even integrated with the iTunes store, which can be easily accessed from the home page.

www.ilounge.com

Next to the Apple Web site, this is the best place to get your information. The iLounge is the unofficial Web site for everything iPod. Here you will find a plethora of articles, message boards, chat forums, product reviews, software downloads, and more. If the topic relates to an iPod, the iLounge is sure to have it.

www.everythingipod.com

You can purchase every iPod model from the Mini to the new iPod Touch on this Web site, right along with all the accessories. This site also hosts tons of useful iPod information.

www.ipodhacks.com

Here is another great Web site powered by a community of iPod fanatics. The iPod Hacks will keep you informed with the latest tips, tricks, and modifications that enable you to take your experience to the next level. This site is user interactive with forums and messaging systems. You can even upload your own iPod tips for others to view.

www.my-ipodaccess.com

Here is your one-stop shop for all iPod downloads. My iPod Access offers a variety of ways to transfer content to your device with more than 300 million songs, movies, music videos, and games. This site offers unrestricted and unlimited downloads.

www.ipoddownloadingpro.com

This Web site has more than 95 million files available for download. iPod Downloading Pro also offers a variety of step-by-step tutorials that enabling you to master your device. With a huge selection of music, movies, and video games, it is easy to understand why this Web site has become a top resource for millions of iPod users.

www.quickipoddownloads.com

This is another great place for downloads. Quick iPod Downloads has a large collection of music, movies, video games, and photos for your device. Movies on this site come available in DVD and Hi-Definition quality.

www.ipodnetdownloads.com

iPod Net Downloads has millions of songs, movies, and TV shows ready to transferred to your iPod. This Web site contains all the tools required to easily convert these files to be used with your iPod. You can also download the latest versions of iTunes and other iPod related software.

www.iPodwebsitereviews.com

This Web site is dedicated to bringing you the best and iPod resources. They compile a list of the most popular iPod Web sites and provide users with reviews on what those particular sites offer.

www.ebay.com

We all know that eBay is the hot spot for nearly all of your needs whether new or old. It also popular for the selling and reselling of iPods. You can find a variety of great deals on eBay for iPods and many handy accessories. Since many of these devices may be previously used, be sure to evaluate a seller's track record before making a full commitment.

www.amazon.com

Amazon is world renowned for hosting numerous products from electronics to novels. They have recently become well known for selling iPod models and accessories. At Amazon, you can find new and used iPods at reasonable prices. Many of these items come with free shipping, a factor you will certainly want to take advantage of.

Bibliography

Personal Reference

My iPod video

iTunes version 7

Web Sites

www.apple.com

www.zune.net

www.real.com

www.rhapsody.com

www.emusic.com

www.musicmatch.com

www.amazon.com

www.griffintechnology.com

http://en.wikipedia.org/wiki/IPod

http://en.wikipedia.org/wiki/ITunes

http://en.wikipedia.org/wiki/Creative_Zen

http://en.wikipedia.org/wiki/Nakamichi

http://en.wikipedia.org/wiki/Rio_%28digital_audio_players%29

Articles

article by Philip Michaels - "2008 Predictions" - MacWorld.com

article by Seth Gilbert - "2008 MacWorld Predictions: Gaming the Odds" - seekingalpha.com

article by Danny Gorog - "MacWorld Predictions for 2008" - apcmag. com

article by Christopher Breen - "Consuming podcasts: Tips and Tricks" - MacWorld.com

article by Jeremy Horwitz - "Cheat Codes - Vortex & Texas Hold Em" - iLounge.com

article by Kirk McElhearn - "Beginner's Guide to Podcast Creation" - iLounge.com article by Kirk McElhearn - "The Complete Guide to iTunes' Podcasts" - **iLounge.com**

article by Kirk McElhearn - "Beginner's Guide to Sharing iTunes Music" - **iLounge.com**

article by Ina Fried - "Finding Harmony among iPod Rivals" - **www.cnet. com**

article by Eric Benderoff - "iPod Rivals Measure up Nicely" - **www. seattletimes.nwsource.com**

article by Chris Robertson - "Key Advantages of buying iPod Products at a Discount Shopping Mall" - **www.ebuyersworld.com**

article by Mathew Honan - "Eight great iPod hacks, mods, and tricks" - **MacWorld.com**

Books

book written by Brad Miser and Tim Robertson - *iPod & iTunes Starter Kit* - 2nd Edition

book written by Kirk McElhearn - *iPod & iTunes Garage*

book written by undisclosed author - *How to Get the Most out of Your iPod* - nicheology.com

Author Biography

Contel Bradford was born and raised in Detroit, Michigan. While he developed a passion for writing at an early age, it was not until the year of 2002 when he realized that writing was his true calling — at that time he completed his first book.

Since then, Contel has self-published four books that include, *Dark Decisions, Dangerously in Love, Thug Nation,* and *Consequences: Crossing The Gentlemen.* He has also published three eBooks with Silks Vault Publishing and has written several screenplays. While his books fall under the urban fiction genre, Contel has a unique style of writing and continues to amaze his readers with original concepts.

As of late, Mr. Bradford has crossed over into nonfiction and has become an active freelance writer. He has produced hundreds of articles and content for a variety of Web sites with topics ranging from Web hosting to content management systems.

Contel is a music lover, something he cannot live without. Though fond of his iPod Video, he is actually more of an iTunes fanatic; he is always creating on the computer. Legendary artists such as Jimi Hendrix, Pink Floyd, Led Zeppelin, Bone Thugs N Harmony, and the No Limit Soldiers get the most play in his iTunes library.

Index